Rabbits: *The Animal Answer Guide*

Rabbits

The Animal Answer Guide

Susan Lumpkin and John Seidensticker

The Johns Hopkins University Press   Baltimore

The Johns Hopkins University Press
2715 North Charles Street
Baltimore, Maryland 21218-4363
www.press.jhu.edu

Library of Congress Cataloging-in-Publication Data

Lumpkin, Susan.
    Rabbits : the animal answer guide / Susan Lumpkin and John Seidensticker.
        p.   cm.
    Includes bibliographical references and index.
    ISBN-13: 978-0-8018-9788-7 (hardcover : alk. paper)
    ISBN-10: 0-8018-9788-2 (hardcover : alk. paper)
    ISBN-13: 978-0-8018-9789-4 (pbk. : alk. paper)
    ISBN-10: 0-8018-9789-0 (pbk. : alk. paper)
    1. Leporidae—Miscellanea. 2. Rabbits—Miscellanea. I. Seidensticker, John.
II. Title.
    QL737.L32L858 2010
    599.32—dc22   2010019455

A catalog record for this book is available from the British Library.

*Special discounts are available for bulk purchases of this book. For more information,
please contact Special Sales at 410-516-6936 or specialsales@press.jhu.edu.*

The Johns Hopkins University Press uses environmentally friendly book
materials, including recycled text paper that is composed of at least 30 percent
post-consumer waste, whenever possible. All of our book papers are acid-free,
and our jackets and covers are printed on paper with recycled content.

And suddenly the rabbit, which had been crouching as if
it were a flower, so still and soft, suddenly burst into life.
Round and round the court it went, as if shot from a gun,
round and round like a furry meteorite, in hard tense circles

D. H. Lawrence

Contents

Acknowledgments xi
Introduction xiii

1 **Introducing Rabbits** 1
What are lagomorphs? 1
What is the difference between rabbits, hares, and pikas? 5
Is this book about rabbits, hares, pikas, or all three? 8
How many kinds of rabbits are there? 8
How are rabbits classified? 12
Why are rabbits important? 15
Why should people care about rabbits? 18
How did rabbits evolve? 19
When did rabbits evolve? 22
What is the oldest fossil rabbit? 25
What were the largest and smallest fossil rabbits? 26

2 **Form and Function of Rabbits** 29
What are the largest and smallest living rabbits? 29
What is the metabolism of a rabbit? 31
Why is rabbit scat round? 33
Why are rabbits always sniffing? 35
Why do rabbits have a "hare lip"? 37
Is it true that rabbit teeth never stop growing? 38
Do rabbits sleep? 39
Can rabbits see color? 40
Do all rabbits have big ears and short tails? 42
Why do rabbits have whiskers? 44
Can rabbits run? 44
Can rabbits swim? 48

3 **Rabbit Colors** 49
Why are so many rabbits brown? 49
What causes the fur colors of rabbits? 54
Do fur colors change in different seasons? 56
What color are a rabbit's eyes? 61
What color are baby rabbits? 61
Are there albino rabbits? 61

4 **Rabbit Behavior** 64
   Are rabbits social? 64
   Do rabbits fight? Do rabbits bite? 70
   How smart are rabbits? 71
   Do rabbits play? 73
   Do rabbits talk? 74
   Who eats rabbits? 79

5 **Rabbit Ecology** 84
   Where do rabbits live? 84
   Where do rabbits sleep? 87
   Do rabbits migrate? 87
   Which geographic regions have the most species of
      rabbits? 88
   Which rabbits have the largest distributions and which the
      most restricted? 92
   How do rabbits survive in the desert? 95
   How do rabbits survive the winter? 98
   Do rabbits hibernate? 99
   Do rabbits get sick? 100
   Are rabbits good for the environment? 104

6 **Reproduction and Development of Rabbits** 111
   How do rabbits reproduce? 111
   How long are female rabbits pregnant? 114
   Where do rabbits give birth? 115
   Do rabbits nest at the same time and in the same place
      every year? 117
   How many babies do rabbits have? 118
   Are all babies in a rabbit's nest full siblings? 120
   Do rabbits care for their young? 122
   How fast do rabbits grow? 126
   How can you tell the age of a rabbit? 128
   How long do rabbits live? 128

7 **Rabbit Foods and Feeding** 131
   What do rabbits eat? 131
   How much do rabbits chew their food? 135
   How do rabbits find food? 137
   Do rabbits drink water? 138
   Do rabbits ever store their food? 141

8   Rabbits and Humans 144
    Do rabbits make good pets? 144
    How were rabbits domesticated? 146
    Why did people say "the rabbit died" to mean a woman
        was pregnant? 149
    Are rabbits used in a lot of experiments? 150
    Do rabbits feel pain? 154
    What if I find a baby rabbit or an injured rabbit or if I hit
        one with my car? 154
    How can I see rabbits in the wild? 156
    Should people feed rabbits? 158

9   Rabbit Problems (from a human viewpoint) 159
    Are rabbits pests? 159
    How can I keep rabbits out of my garden? 169
    Are rabbits bad for lawns? 169
    Are rabbits dangerous? 170
    Is it safe to eat rabbits? 171
    What should I do if I get bitten by a rabbit? 171

10 Human Problems (from a rabbit's viewpoint) 172
    Are any rabbits endangered? 172
    Will climate change affect rabbits? 176
    Are rabbits affected by pollution? 181
    Why do people hunt and eat rabbits? 183
    What products are made from rabbits? 187
    Why do so many rabbits get hit by cars? 190
    Do house cats kill rabbits? 193
    What can an ordinary citizen do to help rabbits? 193

11 Rabbits in Stories and Literature 196
    What roles do rabbits play in mythology and religion? 196
    What roles have rabbits played in language and literature? 200
    What roles do rabbits play in popular culture? 204

12 "Rabbitology" 208
    Who studies rabbits? 208
    Which species are best known? 209
    Which species are least known? 211
    How do scientists tell rabbits apart? 212
    How can I become an expert on rabbits? 213

Contents

Appendix: Rabbits of the World 217
Bibliography 221
Index 229

Contents

Acknowledgments

We dedicate this book to the late Robert S. Hoffmann, who taught John mammalogy, and to the late John F. Eisenberg, who taught Susan to ask interesting questions about mammals.

Many people shared their research, ideas, and time and encouraged us to write this book. We are especially grateful to Néstor Fernández, who showed us European rabbits and Iberian lynx in Doñana National Park, where our interest in rabbits began, and to Miguel Delibes, who arranged for Nestor to host us.

We also thank Lee R. Berger, Howard Cheng, Art Drauglis, John Flux, Tom Friedel, Neil Furey, Ted Grand, Haplochromis, Chris Harshaw, Jeremy Holden, Shah Jahan, Brian Kraatz, Jeff Kerby, Robert K. Lawton, John Litvaitis, Paolo Lombardo, Barney Long, Sriyanie Miththapala, Kate O'Brien, Tamara Rioja Paradela, Silviu Petrovan, Michael J. Plagens, Colin Poole, Janet Rachlow, Mario Sacramento, Jason Schmidt, Walter Siegmund, Andrew Smith, Bill Tietjen, Thomas Voekler, Chris Wemmer, Gehan de Silva Wijeyeratne, Don Wilson, Jim Witham, Fumio Yamada, and the many, many lagomorph biologists whose work is cited throughout the text and in the bibliography.

Finally, we thank our daughter, Lala Seidensticker, for appreciating our enthusiasm and cheerfully tolerating us.

Introduction

A few years ago we walked through Spain's Doñana National Park hoping to catch a glimpse of a very rare cat. Doñana is home to a few of the last Iberian lynx, a beautiful cat many experts believe will be the first felid to go extinct since the sabertooth. We were thrilled to spot one in the dusky light and even follow it for a bit as it set out on an evening hunt. A few minutes later, we were equally excited to see a European rabbit, the lynx's near-exclusive prey and one of the world's most abundant and widespread mammals.

It is an ecological irony, that, thanks to human efforts, these rabbits now occupy most of the rest of Western Europe, Australia, parts of South America, North Africa, and more than 800 islands around the world, but they are disappearing in Iberia, their sole home for many thousands of years. This experience piqued our curiosity about European rabbits and then about their kin, the other 90 or so species of rabbits, hares, and pikas that make up the mammalian order Lagomorpha. Having spent years exploring the world of predators—cats and other carnivores—it was time to take a closer look at some of their most important prey.

Lagomorphs feed many carnivores, including many people. They also fire our imaginations. From the dawn of history, rabbits and hares have figured in stories, symbols, and myths of people around the world, with at least one species native to every continent except Australia and Antarctica. From "the rabbit in the moon" of Asian folklore and hare tricksters in literature from around the globe to Peter Rabbit and Bugs Bunny, rabbits have long loomed large in the human psyche, all the more so for their association with sexuality and fertility, from representing Aphrodite to their personification as Playboy bunnies.

Yet while every small child knows what a rabbit looks like, and hundreds of thousands of Americans keep rabbits as pets, few people know much about lagomorphs as wild animals living outside of cabbage patches. Fewer still know anything about the diverse forms they come in, or of their importance in ecosystems, or that several of their kind are among the most endangered species on Earth. It's difficult to convince people that rabbits are in trouble because, if nothing else, everyone "knows" that rabbits breed like rabbits. We hope our book will help change that misperception.

Apart from this, lagomorphs are fascinating in their own right. In this book, we discuss all aspects of the lagomorphs: what is known and what remains to be learned about the diverse species; their anatomy and physiology,

Critically endangered Iberian lynx (*Lynx pardina*) rely on European rabbits (*Oryctolagus cuniculus*), which are near threatened in Iberia but abundant elsewhere. Lynx photo © John Seidensticker; rabbit photo © Paolo Lombardo

evolutionary history, ecology and behavior, and conservation; and their historic and contemporary relationships with people, as both wild and domesticated animals.

We hope that as readers learn about these animals they will find them as intriguing as we do and will be inspired to join efforts to conserve them.

Rabbits: *The Animal Answer Guide*

Chapter 1

Introducing Rabbits

What are lagomorphs?

What, if anything, is a rabbit?

ALBERT F. WOOD

Rabbits and hares are lagomorphs, a word that means "hare-shaped" in Greek. There are few people who can't readily picture rabbits and hares. These animals are found around the world as native or introduced species and figure prominently in myths and legends, art and literature, cuisine and popular culture. Far less well known than rabbits and hares is the third group of the order Lagomorpha: the pikas. Pikas are small mammals resembling guinea pigs that reside mostly in remote mountains and deserts in parts of Asia; two kinds of pikas are found in North America. All lagomorphs are most notably prey species, with extensive adaptations for evading the many mammals and birds that hunt them.

Many people automatically think that lagomorphs are rodents, lumping them with rats, mice, and other relatively small, scurrying herbivorous creatures. For a long time, scientists shared this misperception. Linnaeus, the father of taxonomy, which is the science of classifying living things, placed both kinds of mammals in the order Rodentia. One feature he identified as uniting these two groups of mammals is gnawing incisors (front teeth) that grow continuously throughout the animals' lives. But Linnaeus also recognized a key difference: Lagomorphs have two extra upper incisors, small peglike teeth that sit behind the prominent front teeth. Thus, the lagomorphs were dubbed the Duplicidentata (duplex-toothed) and the rodents the Simplicidentata (simple-toothed).

Everyone knows what a rabbit looks like. Photo by sevenstar, Wikimedia Commons

As scientists studied rodents and lagomorphs more closely over the years, they found other differences between these groups. For example, lagomorph incisors have a single, unpigmented layer of enamel that surrounds the teeth; rodent incisors have a pigmented double layer of enamel on the front of the teeth. In 1912, James W. Gidley, a paleontologist at the Smithsonian Institution, proposed elevating the lagomorphs to their own, separate order of mammals, based on both dental and limb features. This is where the classification of lagomorphs stands today.

More vexing and contentious has been the question of how lagomorphs are related to the rest of the mammals, an issue that has long puzzled biologists. Gidley believed they were most closely related to artiodactyls, such as bison, deer, antelope, and the like. Renowned paleontologist George Gaylord Simpson reverted to linking rabbits with rodents, although he noted that this was "permitted by our ignorance rather than sustained by our knowledge." Acknowledging the problem, paleontologist Albert F. Wood titled a 1957 scientific paper, "What, If Anything, Is a Rabbit?" A 2004 article by Emmanuel J. P. Douzery and Dorothée Huchon called "Rabbits, If Anything, Are Likely Glires" reflects the current consensus that they are most closely related to rodents—Glires being the name for the group that includes both lagomorphs and rodents.

Rabbits, hares, and pikas are vertebrates (animals with backbones) in

Rabbits: The Animal Answer Guide

the class Mammalia. Like humans and about 5,420 other mammal species, lagomorphs have hair and feed their young with milk produced by mammary glands (this is the origin of the word "mammal"). Scientists currently recognize 29 orders in the class Mammalia. Examples of other orders are Carnivora (cats, dogs, and their relatives) and Primates (monkeys, apes, and people). Like all mammals (and birds), lagomorphs are endotherms, that is, they generate heat internally through a high metabolic rate and usually maintain a fairly high and constant body temperature regardless of ambient temperature, within reasonable limits; this is often called being "warmblooded." Lagomorphs, like all mammals except platypuses and echidnas, give birth to live young. Their basic body plan is typical of mammals and is without any extraordinary adaptations such as those that bats and whales evolved for life in the air and the sea, respectively. All mammals have similar senses—hearing, vision, taste and smell, balance, and touch—although sensory abilities vary among groups and species. Lagomorphs, for instance, have excellent senses of smell and hearing. The 90 or so species of living lagomorphs share a suite of features that differentiate them from the other orders of mammals, including the Rodentia. Morphologically they are all very similar and exhibit a number of primitive mammalian traits coupled with highly evolved teeth, skull, and limb characteristics.

TEETH AND SKULL. In addition to ever-growing incisors and the peg teeth behind them, all lagomorphs have a diastema, or gap, where the canine teeth are in people and many other mammals, between the incisors and the cheek teeth. Rabbits and hares have three upper and two lower premolars and three upper and lower molars on each side of the jaw, for a total of 28 teeth; pikas have one fewer upper molar in each jaw, for a total of 26. When lagomorphs gnaw, their upper and lower incisors meet directly, as rodent incisors do; their lower incisors can also close behind their upper ones, as yours can but not a rodent's. Parts of the skulls of rabbits and hares are not solid bone; instead, they are thin and fenestrated, resembling lace fabric.

LEGS AND FEET; FUR, EARS, AND TAILS. Rabbits and hares have long legs and large hind feet, while pikas' limbs are relatively short. In the rabbits and hares, the tibia and fibula (two bones, corresponding to our shin and calf bones, in the lower portion of the hind limbs) are fused; this adds strength to the limbs and also reduces their weight—important adaptations for running speed. All lagomorphs have five toes with strong claws on the forefeet and four on the hind feet. None uses its front feet like hands to hold or manipulate food or other objects. Dense fur covers the soles of the

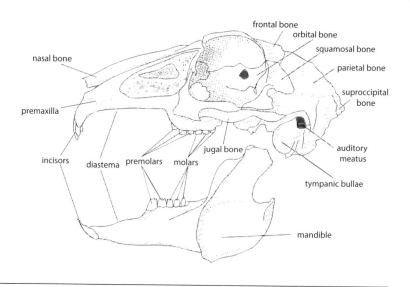

The skull of a rabbit. Courtesy of Bill Tietjen, Bellarmine University

feet; mats of hair replace the fleshy foot pads characteristic of most mammals. Lagomorphs run on their toes, called digitigrade posture, but put the entire sole on the ground when walking slowly or at rest, called plantigrade posture. Lagomorphs have long, soft fur generally in camouflage colors. Their tails are short and round, and those of pikas are nearly invisible (although the American pika's tail is longer relative to its body size than that of any other lagomorph). The ears of rabbits and hares are large and those of pikas, although smaller, are still prominent.

DIGESTIVE SYSTEM. All lagomorphs are herbivores, feeding entirely or almost so on grasses, leaves, bark, seeds, and roots. To get the most out of this low-quality diet, they pass it through their digestive system twice by eating their feces, called "coprophagy" (from the Greek for feces, *copros*, and to eat, *phagein*). Nutrients are extracted from the vegetation they eat through a process of fermentation. All lagomorphs are hindgut fermenters, as opposed to cows and their relatives, which are foregut fermenters. Their guts are large, especially the cecum, a chamber of the gut where fermentation takes place; the digestive system accounts for much of the body mass of pikas (see "Why is rabbit scat round?" in chapter 2).

ODDS AND ENDS. Rabbits and hares have one external opening used for reproduction (copulation and parturition) and for elimination of urine and another one for excreting feces; pikas have a single opening for all three functions. The testicles of males are placed in front of the penis, a trait found in marsupials like kangaroos but in no other placental mammals. The testicles are usually held inside the body unless the animal is breeding, during

Rabbits: The Animal Answer Guide

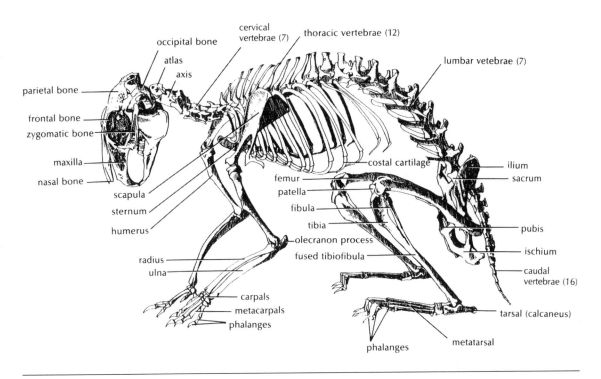

The skeleton of a rabbit. Courtesy Geoffrey Stein; adapted from Bruce D. Wingerd, 1985

which time they descend into the scrotum. Pikas are an exception to this, because they lack a scrotum. Male lagomorphs do not have a baculum—a bone in the penis that is found in many other mammals including most rodents. Female lagomorphs have two separate uteri, each with a cervix, as do rodents. The number of nipples varies with the species, generally ranging from six to eight (three or four pairs on each side of the ventral surface).

What is the difference between rabbits, hares, and pikas?

Let's start with a clarification of terminology. Pikas are always pikas, even when they are referred to as "calling (or whistling) hares" or "mouse hares" or "junior bunnies." Very often, what we call a rabbit is a hare or vice versa. For instance, the black-tailed jackrabbit (*Lepus californicus*), a familiar species of the western United States, is a hare, while the hispid hare, native to parts of India and Nepal, is a rabbit, as reflected in its preferred common name, the bristly rabbit (*Caprolagus hispidus*). Scientists now mostly agree that the word "hare" applies only to the 32 or so species in the genus *Lepus*, which includes species called, correctly, hares as well as the jackrabbits. "Rabbit" describes about 26 species in 10 different genera, with the largest number occupying the genus *Sylvilagus*, the North American cottontails. The most ubiquitous of rabbits, the European rabbit (*Oryctolagus cuniculus*), is the only species in its genus.

"Bunny" is mostly a child's word, applied to either rabbits or hares. (Albert F. Wood suggested using "bunny" as the common name for all lagomorphs but, regrettably, it never caught on.) Bugs Bunny, for instance, is clearly a hare. In fact, Bugs first appeared in 1940 in a film called *A Wild Hare*. Trix cereal's silly rabbit looks more like a hare, too. Also, some animals called rabbits or hares are not lagomorphs. For instance, the Patagonian hare or mara (*Dolichotis patagonum*) is a large rodent, as is the springhare (*Pedetes capensis*) of Africa.

Scientists divide the living lagomorph order into two families: the pikas are from the family Ochotonidae and the rabbits and hares are called leporids and are from the family Leporidae. (A member of a third family, the Prolagidae, the Sardinian pika, *Prolagus sardus*, is now extinct but survived until perhaps the 1700s.) Dividing this order into two families reflects our understanding of their evolutionary relationship: Rabbits and hares are more closely related to each other than they are to pikas.

Pikas are rounder and smaller than leporids. Their ears are petite and round, although large for their body size, their legs are short, and their eyes small. Most pikas are active during the day and are highly vocal. About half of the pika species live on talus (loose rocks) above ground; the rest use burrows. To escape predators, pikas drop out of sight among the rocks or scramble into their burrows.

Hares and rabbits are larger and more angular in shape than pikas. Hares and rabbits have large eyes, adapted to their being mostly nocturnal or crepuscular, active at dawn and dusk. Rabbits and hares also tend to be fairly quiet, with a small repertoire of vocalizations. Hares are generally bigger than rabbits, with longer legs and long ears usually tipped in black fur. Some rabbits use burrows and retreat underground to escape danger or scamper into the shelter of tangled brush, while hares usually run away, at high sustained speeds.

A major difference between rabbits and hares is in their reproductive traits. Rabbits are born in a nest of fur and grass in a burrow or special depression on the ground, called a form. In some species, the rabbits dig their own burrows; in others, they use burrows abandoned by other animals. Baby rabbits are called kittens or sometimes pups. They are altricial (needing a parent's care at birth), being born naked and helpless with their eyes shut. Hare babies, called leverets, are born precocial (largely able to care for themselves): fully furred, open-eyed, and ready to go. As with rabbits, pika newborns are altricial.

Rabbits and hares are generally solitary, although, exceptionally, the

The mountain cottontail is one of about 31 species of rabbits, which are intermediate in size between pikas and hares. Photo © Jim Witham

About 32 species of small, roundish pikas, like this plateau pika, form the lagomorph family Ochotonidae.
Photo © Andrew Smith

European rabbit (*O. cuniculus*) is very social and some hare species sometimes form large herds. Pikas range from highly unsocial in some species to highly social in others. Finally, among these general differences between the groups, hares are adapted to life in open country—plains, tundra, savanna, desert, and steppe, while rabbits occupy diverse habitats from rain forests and marshes to relatively open grassland and deserts.

The tapeti, a widespread cottontail rabbit in South America, may prove to be multiple species. Photo © Mario Sacramento

Is this book about rabbits, hares, pikas, or all three?

This book is about rabbits, hares, and pikas—all of the living lagomorphs. Just to simplify things, though, we'll refer to all three in the questions as rabbits.

How many kinds of rabbits are there?

A species, or kind, of animal consists of a number of individuals with very similar features and genetic makeup. Males and females within a species are able to breed and produce viable young that are fertile. Under typical natural conditions, they do not breed with members of other species, although exceptions occur; interspecific breeding may be common among some hares whose distributions overlap.

According to the latest review (2005) by lagomorph specialists Andrew Young and Robert Hoffmann in Don Wilson and DeeAnn Reeder's authoritative *Mammal Species of the World*, there are 91 species of lagomorphs: 30 species of pikas, 32 species of hares, and 29 species of rabbits. This number is almost certain to change, however, as scientists learn more about them. In fact, the number has changed significantly since the first edition of *Mammal Species* was published in 1982, when just 61 species—18 pikas, 19 hares, and 24 rabbits—were named. The second edition, published in 1993, increased that number to 79: 25 pikas, 30 hares, and 24 rabbits. And even since the 2005 edition, several more species have been added to the list.

What accounts for the rapid increase in species? First, a few new species,

Rabbits: The Animal Answer Guide

previously unknown to scientists, were identified. The Ili pika (*Ochotona iliensis*) was discovered in the Tienshan Mountain of China's Xinjiang province in 1986, and the black pika (*O. nigritia*) was found in China's Yunnan province in 2000. A most sensational scientific find was the discovery of a new rabbit in 1999, named the Annamite striped rabbit (*Nesolagus timminsi*) for the Annamite Mountains that straddle the border of Vietnam and Laos in Southeast Asia. Little explored until the 1990s, these mountains have since rewarded scientists with several other new species, including a large bovid (cowlike animal) with very long horns called the saola (*Pseudoryx nghetinhensis*). A British biologist, Rob Timmins, found three of the rabbits in a Laotian market—unfortunately all were dead. DNA analysis revealed that this species is closely related to the almost equally unknown Sumatran striped rabbit (*N. netscheri*). In fact, DNA from the new rabbits had to be compared with DNA extracted from museum skins of the Sumatran striped rabbit because the last time anyone sighted these elusive creatures in the flesh was in 1972! Since then, its image has been captured a few times in camera traps.

Other additions to the number of lagomorphs came from splitting existing species based on new scientific research, including studies of morphology, chromosomes, and genetics.

Only in 1986 did scientists examine the chromosomes of the New England cottontail (*Sylvilagus transitionalis*) to discover that animals in the southern Appalachians had 46 chromosomes, while those in the rest of the New England cottontail's range had 52. Thus, the southern Appalachian form was named the Appalachian cottontail (*S. obscurus*).

A new cottontail from Venezuela was described in 2000. The Venezuelan lowland rabbit (*S. varynaensis*) was essentially hiding in plain sight: biologists trying to learn more about the two previously known species of cottontails in that country were surprised to find specimens sufficiently different from both to justify a new species. (These are called "cryptic species.") Zoologist Luis Ruedas, who specializes in the study of *Sylvilagus* species, elevated two subspecies of the very wide-ranging eastern cottontail (*S. floridanus*) to full species: the robust cottontail (*S. robustus*) and the Manzano Mountain cottontail (*S. cognatus*). Similarly, he deemed the Panamaian subspecies of the tapeti (*S. brasiliensis*) of Central and South America a full species (*S. gabbi*) in 2007. Others will likely be identified, especially among the eastern cottontail. As Ruedas wrote, "*Sylvilagus floridanus* can not be allowed to exist as is currently understood," because this species name now probably includes several different species. The same could be said about the tapeti.

Among Africa's rock rabbits, genus *Pronolagus*, molecular genetic studies—an analysis of the differences in DNA—recently forced splitting one of the four named species into two, and other analyses suggest there is likely a third species separate from these two.

The tapeti, a widespread cottontail rabbit in South America, may prove to be multiple species. Photo © Mario Sacramento

Most challenging and contentious is the number of hares, genus *Lepus*. All hares examined have 48 chromosomes (unlike humans, who have 46), so that's no help. Because several apparent species have very wide ranges, morphological comparisons are hampered by variations in size, color, and other physical features due to adaptations to local conditions. The distributions of some species have expanded and contracted in historical as well as evolutionary time. Further, people have moved hares from place to place over the past 3,000 years. Finally, the hare radiation—diversification into multiple species—is fairly recent in evolutionary time. Speciation into groups divided among North America, Europe, Asia, and Africa took place between about 3 and 6 million years ago, while further speciation within regions occurred less than 800,000 years ago.

Almost certainly, the known number of lagomorph species will continue to grow, not only because this is a relatively little-studied group but also because the number of named mammal species in general is growing. Some scientists predict that ultimately the number of mammals will reach about 7,500, about half again those named today! As in the lagomorphs, continued exploration of little-explored parts of the Earth, such as the Annamite Mountains mentioned above, will account for some of this increase. But increasingly sophisticated analyses to detect genetic differences between populations will likely account for most of it. It is becoming clear that even where there are few or very small morphological differences between animals in different populations there may be genetic differences, even where their ranges broadly overlap and even if animals from the two (or more) populations sometime interbreed.

Despite the growing number of described species, there are still rela-

Rabbits: The Animal Answer Guide

tively few lagomorph species compared with rodents, which number more than 2,000 species. The reasons for this are obscure.

The evolution of species within a region depends on the geographical configuration of the region, spatial arrangements of habitats, and the ability of the species to disperse. The process of speciation, or the formation of a new species, is the full sequence of events leading to the splitting of one population of organisms into two or more populations that become reproductively isolated from each other. A new species may arise when the parental species is divided by a geographic, vegetative, or other extrinsic barrier. This is called "allopatric," or geographical, speciation. The flooding of the Bering Strait between Asia and Alaska and western Canada when glaciers began to melt end of the Pleistocene probably separated the North American pikas from the Asian ones. The diversity of pika species in Asia is related to the uplifting of the Tibetan Plateau from 3.4 million to 1.6 million years ago, which isolated populations from one another. Advancing glaciers earlier in the Pleistocene also pushed widespread lagomorph populations into isolated ice-free refuges; the European rabbit took refuge in the Iberian Peninsula for instance, while the European hare (*Lepus europeaus*) may have retreated there as well as to the southern Balkans.

A new species may also arise through the modification of a peripherally isolated founder population, which is another kind of allopatric speciation. The founders may be just a few individuals or even a single pregnant female. Some of these founder populations go extinct; some may eventually merge again with the parent species; some become "good" species, or true species that retain any distinguishing characteristics over time. A founder population may find itself in an entirely new environment or a changing environment that offers the ideal situation for evolutionary departures into new niches and adaptive zones. The cottontails may be an example, with species adapted to habitats ranging from dry desert to tropical rain forest as they moved through North and South America.

There are large gaps in the current distributions of some related lagomorph species, such as the striped rabbits that are found only in Sumatra, an Indonesian island, and the Annamite Mountains of Vietnam and Laos. Similarly, the Corsican hare (*L. corsicanus*), found in central and southern Italy and Sicily, is closely related to the broom hare (*L. castroviejoi*) of the Iberian Peninsula, suggesting that their common ancestor had a wider distribution in southern Europe. This may be because the species became extinct in parts of a once-continuous range. It is also possible for members of a species to establish founder populations far away after dispersing across unsuitable terrain such as water, mountains, or inhospitable habitat. This is called "vicariant speciation." These widely dispersed populations may eventually become full species themselves. A recent study by evolutionary geneticist Conrad A. Matthee and his colleagues examined the genetics of leporids to determine how, where, and when they diversified. Their results suggest that the current leporid distribution is the result of at least nine different events of speciation through dispersal and vicariance.

"Sympatric speciation," or the splitting of a single breeding population into more than one species without geographical or other barrier between them, may occur but there are no good examples among lagomorphs or even among mammals in general.

How are rabbits classified?

RABBITS. The rabbits are divided into 10 genera, most of which include only one species; the most diverse genus is *Sylvilagus*, with at least 17 and perhaps up to 30 species. Most of the rabbits have small to extremely small natural ranges, although thanks to human intervention, the European rabbit now has one of the largest distributions of any mammal.

Three genera of Western Hemisphere rabbits are endemic, or native, to a region.

Romerolagus: There is only one species in this genus, *R. diazi*, the volcano rabbit of Mexico. This small rabbit averages about 0.45 kilogram (1 pound) in weight and 9 to 12 centimeters (9 to 12 inches) in length, and, like pikas, has no visible tail. This species is found only in the central part of Mexico, in volcanic mountains not far from Mexico City. Living as high as 4,250 meters (14,000 feet), it inhabits pine forests with rocky substrates and dense grassy undergrowth; grasses also form the mainstay of its diet. In Mexico, it is called the zacatuche, or grass rabbit. Genetic analyses suggest that this is a primitive rabbit, perhaps the most primitive of the living leporids. In one view, a North America leporid species diverged about 13 million years ago into a lineage giving rise to the volcano rabbit and another giving rise to hares, cottontails, the pygmy rabbit, and some of the other rabbits.

Brachylagus: The pygmy rabbit (*B. idahoensis*) is the smallest of rabbits and it, too, is the only living member of its genus. It is probably most closely related to cottontails. This species is found primarily in the dry Great Basin and adjacent mountains of the western United States, with a separate population, recently extinct in the wild, in southeastern Washington state. Its precise distribution largely matches that of big sagebrush (*Artemisia tridentata*). The leaves of these large plants stay green all year and provide the rabbit with almost all of its winter food.

Sylvilagus: The members of this genus are all very similar morphologically and to some extent behaviorally, although there is much chromosomal variability and differences in some reproductive parameters, such as litter size and number of litters per year. One species or another live throughout the United States and barely into southern Canada, most of Mexico and Central America, and parts of northern South America. Most appear to be fairly adaptable in terms of their diet, and may live in tropical, deciduous, coniferous, and temperate forests, cool and warm deserts, in riparian habitats, and in agricultural and other human-dominated landscapes, from sea level to elevations of more than 3,000 meters (10,000 feet). But many are habitat specialists and live only within restricted areas within these broad habitat types that meet their particular needs.

Africa also has three endemic genera of rabbits.

Poelagus: This is another monotypic (single species) genus, found in Africa. *P. marjorita*, the Bunyoro rabbit, is named for the place in Uganda where it was first described. It also ranges into savanna-woodland habitats in central Africa; a separate population lives in Angola. Not much is known about this species and there has been little agreement about what might be its closest relatives. The Matthee analysis showed, however, that its sister taxa may be the African rock rabbits and that these two genera share a common ancestor with Asia's striped rabbits.

Bunolagus: Found only in the Karoo Desert region of Cape Province, South Africa—and there only along a single river system—the riverine rabbit (*B. monticularis*) has particularly long ears. Unusual among typically prolific rabbits, a female riverine rabbit gives birth to a single offspring per year. Its relationship to other rabbits is also obscure, but it may be a sister taxon to the European rabbit, with these two sharing a common ancestor with one that split into the bristly rabbit and the Amami rabbit.

Pronolagus: The four or five species of *Pronolagus* are known as rock rabbits. They are native to sub-Saharan Africa but have relatively small, patchy distributions there and are confined to rocky, broken habitat with bushes and grasses for cover. All have reddish fur on their legs and red or dark tails.

Three genera are also endemic to Asia.

Pentalagus: The Amami rabbit (*P. furnessi*) is believed to be a primitive rabbit whose closest relatives, depending on the analysis, may be Africa's riverine rabbit or Mexico's volcano rabbit. A medium-sized rabbit, it lives in moist subtropical forest on only two islands in southern Japan's Ryukyu Archipelago. The Amami rabbit is social and uses an extensive complex burrow system, a behavior it shares with the European rabbit among the leporids.

Caprolagus: The bristly rabbit (*C. hispidus*) lives in tall grasslands at the base of the Himalayas in Assam in India, Nepal, and Bhutan but once was found in this habitat throughout India and into Bangladesh. An outer layer of bristly fur over a short, fine layer of fur gives this dark brown and black rabbit its name.

Nesolagus: Two species, one recently discovered, form this genus of extremely poorly known rabbits. They differ from all other rabbits in being striped, hence their names, the Sumatran striped rabbit (*N. netscheri*) and the Annamite striped rabbit (*N. timminsi*). They have relatively small ears. Both species live in moist tropical forest. Several studies suggest their nearest relatives are the Bunyoro rabbit and the rock rabbits of Africa.

Only one genus is endemic to Europe.

Oryctolagus: The European rabbit (*O. cuniculus*) is today found around the globe, but its recent natural range was confined to the Iberian Peninsula in southwestern Europe and a bit of northern Africa and southern France. These rabbits are the last of several species of *Oryctolagus* that once lived more widely in Europe but were been forced into their southern refugia (an area that remains unaltered in times of dramatic climate change) during the glaciations of the Pleistocene. They now live throughout much of Western Europe and have been introduced around the globe. Genetic analyses by Nuno Ferrand and his colleagues and others show that European rabbits form two distinct populations. One, *O. c. algirus*, is found in the southwestern portion of the Iberian Peninsula; the other, *O. c. cuniculus*, in the northeast Iberian Peninsula, France, and the rest of the world. There is a narrow zone of hybridization where the distribution of the two populations meet, but the split between the two is ancient. *O. c. cuniculus* is also ancestral to all domestic rabbits. As noted above, the European rabbit is one of only two leporids to shelter in social groups in large, complex burrow systems.

HARES. The about 32 species of hares all belong to the genus *Lepus*. Considered an "advanced" genus based on physical characters, hares radiated into today's diverse species more recently than the rabbits did, although the genus may have arisen much earlier. Hares owe their success in grassland habitats to adaptations for high-speed running to escape predators, rapid reproductive rates, and highly precocial young. Hares occur in grassland habitats throughout Eurasia, Africa, and North America but not naturally in South America. In the Western Hemisphere, the southernmost hare, the Tehuantepec jackrabbit (*Lepus flavigularis*), reaches only as far south as southern Mexico. European hares were introduced into South America, with the greatest success in Chile and Argentina, which both have large populations, but they are found in most of that continent. European hares have also been introduced to eastern North America, New Zealand, and Australia.

PIKAS. The 30 living species of pikas, all in the genus *Ochotona*, are small and more egg-shaped than the rabbits and hares. The pikas are restricted to cold climates and most live at high to very high elevations. Most of the species are found in East Asia, with 24 species found in China. Two species live in North America, and one is found as far west as the eastern edge of Europe. The extinct Sardinian pika is often mentioned in lists of the "living" lagomorphs because it became extinct relatively recently.

Why are rabbits important?

If you had to use just one word to characterize lagomorphs, it would be prey, or food. Chinese legend has it that the moon is inhabited by a white hare. The explanation goes something like this: Three wise men took on the guise of old beggars and asked a fox, a monkey, and a hare for something to eat. Even though they had food to offer, the fox and monkey declined to share. The hare had nothing to give so it leapt into a fire, cooking itself to make a meal for the poor old men. The animal was rewarded for its sacrifice with an exalted position in the heavens.

The biological truth in this story is that seemingly everyone eats lagomorphs. In the Iberian Peninsula (Spain and Portugal), biologists report about 40 different species that eat European rabbits. This list contains 4 reptiles, 19 birds of prey, and 17 mammals and includes the endangered Spanish imperial eagle (*Aquila heliaca*)—a huge raptor with a wing span of about 2 meters (7 feet) and the critically endangered Iberian lynx (*Lynx pardina*), which eats almost nothing but European rabbits. Between 75 and 95% of the Iberian lynx's dietary intake is rabbit, and a lynx's daily energy needs can be satisfied by a single, 1 kilogram (2.2 pound) bunny. Declining rabbit numbers, largely the result of disease and habitat conversion, are believed to be directly responsible for the lynx's critical status.

Some other predators also depend nearly exclusively on lagomorphs. In North America, the Canada lynx (*L. canadensis*) specializes on eating snowshoe hares (*L. americanus*). Naturalist Ernest Thompson Seton wrote of the Canada lynx: "It lives on rabbits, follows the rabbits, thinks rabbits, increases with them, and on their failure dies of starvation in the unrabbited woods." Bobcats (*L. rufus*) specialize on cottontail rabbits. In Tibet, brown bears (*Ursus arctos*) live on plateau pikas (*Ochotona curzoniae*), so much so that the Tibetan bears were first described as a species, *U. lagomyiarius*, or bear pika-eater.

Lagomorphs form some portion of the diets of most cat species; wolves and foxes; weasels, ferrets, and other mustelids; many birds of prey such as eagles and owls; and even snakes, either as mainstays or as fallback food when other prey are scarce. Surprisingly, in the North American boreal forest, red squirrels and arctic ground squirrels (*Tamiasciurus hudsonicus* and *Spermophilus parryii*), which are generally seed-eaters, eat a large number

Bobcats (*Lynx rufus*) specialize in hunting cottontail rabbits. Photo by John Jarvis, USFWS

of snowshoe hare babies under 4 weeks of age, when the leverets are still spending most of their time in the nest.

Even where European rabbits are relative newcomers to the menu of prey species, they play an important role in maintaining predators. In Australia, where rabbits were introduced about 130 years ago, three species of introduced mammal—foxes (*Vulpes vulpes*), cats (*Felis catus*), and dingos (*Canis familiaris dingo*)—eat rabbits almost exclusively on farmland where native prey of this size are gone. In addition, 13 species of birds of prey depend on rabbits. Where other prey are scarce, the wedge-tailed eagle's (*A. audax*) diet may be 97% rabbit. After rabbits were decimated by rabbit hemorrhagic disease in the mid-1990s (see "Do rabbits get sick?" in chapter 5), wedge-tailed eagles failed to breed for 3 years in a row in one region of the continent.

At various time and places, lagomorphs have been critical parts of human diets as well. Archeologists have evidence of people hunting rabbits in the south of France 120,000 years ago; undoubtedly, hominids were eating lagomorphs earlier than that. Some archeologists have shown that rabbits and hares became very important in the diets of people in southern Europe

Rabbits: The Animal Answer Guide

Usually vegetarians, red squirrels prey heavily on baby snowshoe hares in the North American boreal forest. Photo by John Jarvis, USFWS

and northern Africa at the end of the Pleistocene, 8,000 to 12,000 years ago, as preferred large game such as antelope and deer became scarcer due to overhunting. Others have argued that lagomorphs were particularly important to improving the nutrition of women and children once the development of technology such as nets and snares enabled them to be caught easily, compared with arduous big-game hunting that was believed to be generally the exclusive province of men.

Archeological research shows that the Anasazi people of the southwestern United States relied heavily on cottontails and jackrabbits for subsistence. Large numbers of these lagomorphs, especially jackrabbits, were rounded up in "drives." The more easily caught cottontails were also taken when people chanced upon them as the cottontails foraged in agricultural plots. Evidence from some sites in the North American boreal forest suggests that snowshoe and other hares figured prominently in the diets of Native Americans living there.

Food is undoubtedly the reason that people have moved rabbits and hares from place to place throughout history. European rabbits were moved from the Iberian Peninsula to islands in the Mediterranean as early as 1400 to 1300 BCE, and the first introductions of European hares to Mediterranean islands may date to 3000 BCE.

During the age of exploration, sailors released domestic European rabbits, as well as pigs and goats, on far-flung oceanic islands so that crews on passing ships could stop for a bit of fresh meat. Men on ships hunting for seals continued this practice into the nineteenth century. Long-time rabbit specialists John E. C. Flux and Peter J. Fullagar documented rabbit introductions to more than 800 islands around the world.

Introducing Rabbits

Today, the United Nations Food and Agriculture Organization and other organizations encourage villagers in developing countries to raise domestic rabbits for food and income.

Apart from their direct role as food for many carnivores, some lagomorphs have other important roles in ecosystems, so much so that they are called "keystone species" (see "Are rabbits good for the environment?" in chapter 5).

Of course, lagomorphs are important in their own right, as fascinating creatures whose evolution, ecology, and behavior provide insights into how the natural world works. Zoologist Jonathan Kingdon wrote, "Each species is a unique realization of the possibilities of being."

Why should people care about rabbits?

There are many reasons that people should care about rabbits in particular and wildlife in general. Conservationists identify four arguments that give value to the survival of wild animals such as rabbits. Two of these arguments justify the value of animals with respect to their contribution to human economic prosperity and survival. These are the practical reasons for conservation.

The first is the utilitarian justification: I want to save rabbits because they provide me with income or some other thing of value. Much of the impetus for the conservation of European rabbits and hares, for instance, comes from hunters' concerns about their declining numbers. Rabbit and hare hunting is a very popular recreational activity and in some places is important to human food security.

Second is the ecological justification: I want to save rabbits because rabbits are good for the environment. We will discuss in other sections the important roles lagomorphs play in their various native environments. More generally, we can turn to the words of Indira Gandhi, who, as India's prime minister from 1966 to 1977 and 1980 to 1984, championed wildlife conservation. She said in a speech, "An environment in which animals and plants become extinct is not safe for human beings either." Thriving populations of rabbits may be an indicator of environmental health and, conversely, a declining population is a symptom of a problem that could ultimately affect humans. The decline of some pikas, for instance, is probably a sign of advancing global climate change.

The third argument involves the value of animals' aesthetic appeal: rabbits are valuable because they are cute or beautiful, inspiring stories and poetry and exciting our imaginations. This is comparable to why we preserve great works of art, and for some people, no more reason than this is required to value rabbits.

Finally, some people believe that rabbits have a right to exist, just as people have certain rights, and we as moral beings, are obliged to ensure their survival.

None of these arguments is better than another. People in different times, places, and situations have different needs and values; some people will find all these arguments compelling. Conservationists invoke all of them in their efforts to protect lagomorphs and other wildlife.

How did rabbits evolve?

This is an exciting period in the world of scientists who study mammalian systematics, or the attempt to determine how mammals evolved and diversified over time and how they are related. In this study, biologists are guided by the theory of common descent and evolution through natural selection. Traditionally, they look at morphology—the basic body plan and diagnostic characteristics such as teeth and bones—of the mammals alive today and trace them back to the first appearance in the fossil record of an animal with these characteristics; it is like starting with the tips of the branches of a tree and working down to the trunk. More recently, increasingly sophisticated studies of genetics have provided new insights into relationships that are sometimes at odds with those inferred from morphology.

Using detailed genetic analyses, scientists have developed a clear and sometimes surprising picture of the evolutionary relationships among mammalian orders. There are four major groups of living placental mammals (excluding marsupials and monotremes); these are called "crown clades." A clade is any group of organisms that includes the most recent common ancestor of all those organisms and all the descendants of that common ancestor. Think of a clade as a dynasty, with a founding ancestor and all of its heirs. The four mammalian crown clades are

- The Xenarthra, which means "strange joints." The living members are anteaters, sloths, and armadillos, all confined to the Americas.
- The Afrotheria (African mammals; -*theria* means mammal), which includes today's tenrecs, golden moles, elephant-shrews, aardvarks, elephants, sea cows, and hyraxes. This clade originated in Africa, but later some groups migrated beyond that continent.
- The Laurasiatheria is named for Laurasia, the northern hemisphere supercontinent formed of North America, Europe, and Asia. This clade includes today's shrews, moles, hedgehogs, bats, whales and dolphins, even- and odd-toed ungulates, carnivores such as cats and dogs, and pangolins.

- The Euarchontoglires (a tongue-twister with a convoluted meaning) includes rabbits, rodents, flying lemurs, tree shrews, and primates. Yes, you can count rabbits as among your nearer mammalian relatives!

Scientists are still debating about which of these clades arose first. Most genetic studies award that honor to either the Xenarthra or the Afrotheria. But a comprehensive examination of a host of fossils published in 2007 turned this on its head, putting the Laurasiatheria and Euarchontoglires as the earliest clades and suggesting that the earliest placental mammal was a rabbit-like creature from Asia. This debate will likely continue as each new study helps to clarify this complicated picture.

Among the Euarchontoglires, lagomorphs and rodents form the clade Glires; these two orders are more closely related to each other than to the other members of this group. The living lagomorphs are divided into two families: the Ochotonidae and the Leporidae. All living pikas belong to the Ochotonidae and to the same genus *Ochotona*.

The relationships among the pikas are uncertain, with a variety of studies dividing the pikas into two or more subgenera based largely on morphology but with little agreement between the authors of these studies. A recent molecular analysis, led by Chinese scientist Ning Yu, which included samples from 27 of the 30 species, suggests that pikas fall in five major species groups that are largely separated geographically. The two North American species are linked to what these authors called the northern group, which includes the alpine pika, the northern pika, and Pallas's pika (*O. alpine*, *O. hyperborea*, and *O. pallasi*). The ancestor of the North America species is believed to have migrated from Asia across Beringia about 2 million years ago, while the Bering Strait was above sea level. The steppe pika (*O. pusilla*) of central Asia forms a group all its own; this species once had a much larger Eurasian distribution—fossils place the steppe pika in Britain during the Pleistocene—and is considered an ancient species. Another pika from the central mountains of China also forms a single-species group. The remaining species are roughly divided into those from the Qinghai-Tibet Plateau and those from the area surrounding the plateau.

How the various cottontails in the New World genus *Sylvilagus* are related remains uncertain, an uncertainty that is compounded by the likelihood that the currently described 17 species are likely to reach as many as 30 when more is learned about them. This is especially true of species in Central and South America ascribed to the very widespread eastern cottontails and to the tapeti, which are each divided into upward of 20 subspecies. Close links described among some "sister taxa," however, may stand. The marsh and swamp rabbits (*S. palustris* and *S. aquaticus*) are closely related

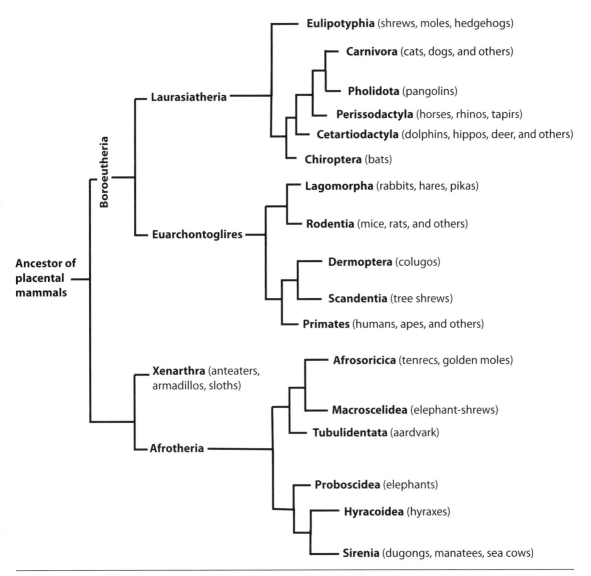

One view of mammal phylogeny; which of Xenarthra, Afrotheria, and Laurasiatheria diverged earliest is still unknown.

and perhaps ancestral to all of other living cottontails. The desert and the mountain cottontails (*S. auduboni* and *S. nutallii*) also shared a common ancestor, as did the Appalachian and the eastern cottontails.

Relationships among hares remain somewhat hazy, too, for the reasons noted above. What's more, there appears to be a surprising amount of hybridization between hares where their distributions meet. Still further, mountain hares (*L. timidus*), which today are largely confined to northern Eurasia, have left traces of their DNA all over the place, according to research by Paulo Celio Alves, Nuno Ferrand, and Franz Suchentrunk. During Pleistocene glaciations, this species expanded its range southward,

Classification of the Lagomorphs

Family	Genus	Number of species
Ochotonidae (pikas)	*Ochotona*	32
Leporidae		
pygmy rabbit	*Brachylagus*	1
riverine rabbit	*Bunolagus*	1
bristly rabbit (hispid hare)	*Caprolagus*	1
European rabbit	*Oryctolagus*	1
striped rabbits	*Nesolagus*	2
Amami rabbit	*Pentalagus*	1
Bunyoro rabbit	*Poelagus*	1
rock rabbits	*Pronolagus*	4
cottontail rabbits	*Sylvilagus*	18
hares	*Lepus*	30

reaching the Iberian Peninsula and leaving behind signs of its presence in all three of the hares still there: the European hare, the broom hare, and the Granada hare (*L. granatensis*).

Some recent genetic analyses, such as that by Matthee mentioned above, have also suggested some surprising affinities. The white-tailed jackrabbit (*L. townsendii*) of the western United States, for instance, clusters most closely not with the other American jackrabbits but with the mountain hare of Eurasia and with the arctic and Alaskan hares (*L. arcticus* and *L. othus*) of far northern North America, although this could be due to the mountain hares' propensity for hybridization. Snowshoe hares appear to be the oldest living hares.

Relationships among the rest of the rabbits, as they are now understood, are described in the next section. Please take almost none of this as established fact, however, as scientists continue to refine their techniques for teasing apart the lagomorphs' fascinating evolutionary history.

When did rabbits evolve?

When the Glires and other crown clades of placental mammals first appeared is highly debated. Based on fossil evidence, many paleontologists argue that the Glires originated close to what is known as the Cretaceous-Tertiary boundary, a period about 65 million years ago that coincided with the extinction of nonavian dinosaurs. In accord with this, the idea is that the demise of the dinosaurs freed up ecological niches that allowed mammals to diversify and flourish, with significant radiations between 65 and 55 million years ago. However, many estimates based on a molecular genetic clock suggest a much earlier appearance of the Glires, with dates ranging from more than 100 million to 80 million years ago. A recent, comprehen-

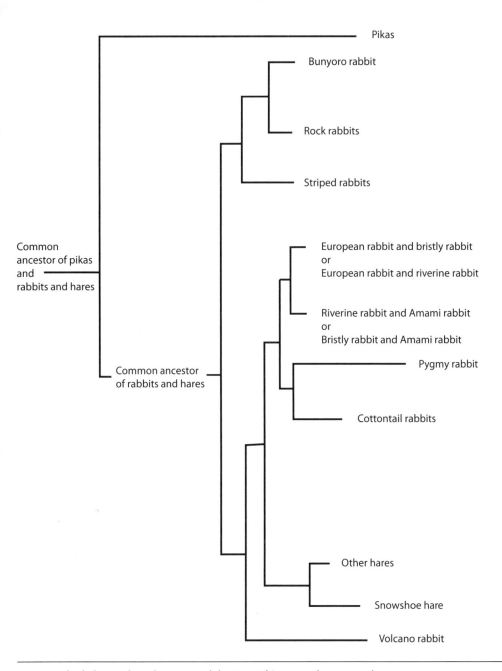

Pikas

Bunyoro rabbit

Rock rabbits

Striped rabbits

European rabbit and bristly rabbit
or
European rabbit and riverine rabbit

Riverine rabbit and Amami rabbit
or
Bristly rabbit and Amami rabbit

Pygmy rabbit

Cottontail rabbits

Other hares

Snowshoe hare

Volcano rabbit

Common ancestor of pikas and rabbits and hares

Common ancestor of rabbits and hares

Lagomorph phylogeny based on research by T. J. Robinson and C. A. Matthee, 2005.

sive analysis suggested that the Glires arose about 92 million years ago and that the split between rodents and lagomorphs occurred as early as about 85 million years ago.

However this debate is resolved, either through the discovery of more fossils or improvements in the calibration of the molecular clock, it is clear from the fossil record that lagomorphs existed at least as long ago as 53 million years before present (see "What is the oldest fossil rabbit?" later in the chapter) and, based on the molecular clock, are most likely considerably older.

Other recent fossil finds in Mongolia date the first known lagomorph relatives, belonging to an extinct group called the Mimotonidae, to about 55 million years ago. These animals indisputably belong to the Duplicidentata, or lagomorph, branch of the Glires, based on their double set of upper incisors and other features. Mimotonids, though, also retained a second, reduced pair of lower incisors that distinguished them from lagomorphs. A nearly complete skeleton of one of these, called *Gomphos elkema*, was described in 2005 by paleontologist Michael J. Novacek and his colleagues. This rabbit-sized creature had the much longer hind limbs than forelimbs and very long feet typical of modern rabbits, but it had more rodent-like tails, molar teeth, and primitive jaws.

Even more compelling evidence that the mimotonids gave rise to the lagomorphs comes from another fossil species from Mongolia described by Brian P. Kraatz in 2009, *Gomphos ellae*. One of the diagnostic features of the lagomorphs, mentioned above, is the diastema, the gap between the incisors and the cheek teeth. There is an evolutionary trend toward greater relative diastema length in the lagomorphs: pikas have short diastemas, comparable to those of the earliest lagomorph fossils. Those of the leporids are longer, but the diastemas of the most primitive leporids are shorter than those of more recent leporids such as hares. The diastema length of *G. elkema* is much shorter than even the most primitive fossil lagomorphs (and that of an early mimotonid shorter still). Until this new fossil was found, there was a large gap (no pun intended) between the two groups. *Gomphos ellae* bridged that gap, with a diastema length intermediate between *G. elkema* and all living and extinct lagomorphs. From these roots (no pun intended again), a large number of lagomorphs arose, with about 78 genera and more than 230 species described from the fossil record.

Ancestral pikas and ancestral rabbits and hares may have diverged about 50 million years ago (see the next question). By about 45 million years ago, the fossil record indicates that lagomorphs were present throughout Eurasia and North America. These early lagomorphs lived until about 10 million years ago. The modern families appeared about 34 million years ago, the Ochotonidae in Asia and the Leporidae in North America (or perhaps Asia), where they eventually replaced the earlier lagomorphs. Gradually members of the modern families colonized all of North America, Eurasia, and Africa. Genetic analysis indicates that the modern genera of rabbits and hare originated between about 13 and 9 million years ago, however there is no fossil evidence of the genera to support this. The cottontails, the only lagomorphs to naturally reach South America, likely originated in

Middle America, diverged after about 5 million years ago, and did not enter South America until after about 4 million years ago.

Between about 7 to 5 million years ago, grasslands came into prominence, replacing forests around the globe. Today, about a third of the Earth's land cover is some sort of grass-dominated habitat. Hares took advantage of this, evolving from a cover-dependent rabbit-like ancestor into open-country dwellers. During this same time period, changing climate led to the formation of a vast ice sheet in Antarctica. This tied up enough water to lower sea levels, which exposed land bridges between once-islanded landmasses. Among other locations, this united Eurasia and North America at the Bering land bridge between what is now Alaska and the Russian Far East. Hares appear to have dispersed both ways, some leaving North America for Eurasia and then into Africa, and others leaving Asia for North America. An earlier period of intercontinental exchange among leporids was also facilitated by a Bering land bridge about 15 million years ago, following 10 or 11 million years of isolation when global ice volume was low and sea levels were thus high.

What is the oldest fossil rabbit?

The oldest fossil lagomorphs have only recently come to light. In 2007, a team led by Kenneth Rose of the Johns Hopkins University described foot bones collected in west-central India as typically lagomorph-like. Dated to the Early Eocene, about 53 million years ago, the fossils are of two ankle bones of a pika-sized animal: the calcaneus, or heel bone, and the talus, the lower part of the ankle joint. Variation in the size and shape of these bones among living lagomorphs is related to the degree of cursoriality (running ability) in the species. That is, in pikas they are shorter and more robust than in most rabbits and hares, where they are long and slender. The fossil ankle bones exhibit a mosaic of features: some more primitive features closely resemble those of living pikas and other, more specialized features are similar to those of leporids. The calcaneus fossils proved to be very similar to those of lagomorph fossils from China that dated to about 48 million years ago, but the Chinese fossils showed differentiation into two forms, one called robust—that is, more pika-like—and the other called gracile—more leporid-like. Together, these fossils suggest that the divergence between the two families of Lagomorphs may have begun about 50 million years ago.

What were the largest and smallest fossil rabbits?

The living lagomorphs display a narrow range of body sizes, from about 100 grams (3.5 ounces) or a bit less for the smallest pikas to around 7 kilograms (15 pounds) for the largest hares. In a book published in 2008, renowned Brown University paleontologist Christine M. Janis remarked, "No significantly larger lagomorphs are known from the fossil record." There is an exception, however, that was discovered shortly after Janis wrote those words. Fossils of a giant rabbit, which has not yet been bestowed with a scientific name, have been unearthed on the Mediterranean island of Menorca, off the coast of Spain. This behemoth's weight is estimated at 14 kilograms (31 pounds), and even apart from its size it was one strange rabbit. It had very small eyes and tympanic bullae (bones in the skull associated with hearing), which hint that it had relatively poor eyesight and hearing. Its spine and legs were relatively short and the plantigrade feet were both short and wide, indicating the giant rabbit was not a speedy runner. Rather, its locomotion was probably slow and lumbering.

What accounts for the evolution of this freakish rabbit is a fascinating tale, reviewed in 2008 by Pere Bover, Josep Quintana, and Josep Antoni Alcover. Menorca is one of the two main islands of the Gymnesic Islands, a subset of the Balearic Islands; the other is Mallorca. These are the most isolated of Mediterranean islands; as a result, they never hosted more than handful of land vertebrates. But their isolation was breached during what is called the Messinian salinity crisis. Discovered only in the early 1970s, the crisis began nearly 6 million years ago (5.96 million years ago, to be precise). At that time, some tectonic event closed the connection between the Mediterranean Sea and the Atlantic Ocean, which now exists at the Straits of Gibraltar between southern Spain and Morocco. Without an influx of water from the ocean, the Mediterranean essentially dried up, and this enabled a few new land animals to reach the far-flung Gymnesics. Then, 5.33 million years ago, the Straits of Gibraltar opened up again. Atlantic waters flooded into the Mediterranean, and the Gymnesics were isolated once again, both from the mainland and from each other. Subsequently, different groups of animals evolved on the two islands.

On Menorca, the giant rabbit dominated. There were no land predators and only two species of owls that may have potentially preyed on the giant rabbit's presumably normal-sized immigrant ancestors, probably in the extinct leporid genus *Alilepus*, which appear to have resembled hares or cottontails. With few aerial predators and no land predators, the rabbits lost their long legs and feet and other adaptations for speed as well as possibly their good eyesight and hearing, adaptations for detecting predators.

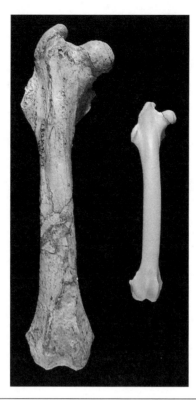

The femur (thigh bone) of an extinct giant rabbit from the island of Menorca and one from a modern-day European rabbit. Photo by SA Monea, Wikimedia Commons / PD

What's more, their large size may have put them out of the reach of the owls or at least made them harder for the owls to capture.

In contrast, a strange, very small, goatlike bovid called the Balearic Islands cave goat (*Myotragus balearicus*) evolved on nearby Mallorca. Again, with no predators but a few owls, these herbivores lost adaptations for speed and predator detection. For instance, unlike all other bovids, their eyes faced forward rather than off to the sides, reducing the wide visual field that lets bovids, and rabbits, see predators approaching from all around. Unlike the rabbits that became giants, however, these animals were dwarfs, likely due to the island's limited food supply. This follows a general evolutionary trend that large mammals get smaller on islands and small ones get larger.

The giant rabbit on Menorca died out about 2 million years ago, when cooler climates led to glaciations, lowering sea levels in Mediterranean and linking Menorca and Mallorca for a time. *Myotragus* invaded Menorca and likely outcompeted the giant rabbit for food. *Myotragus* survived until about 3600 to 2000 BCE, when it is likely that the first people to populate the islands wiped them out.

The largest known fossil pikas (*Gymnesicolagus*) were also found on the Gymnesic Islands, where they went extinct before the Messinian salinity crisis. These pikas reached the size of a large modern hare, about 5.5 kilograms (12 pounds).

No one has identified the smallest fossil lagomorphs, but there are a variety of fossils a bit smaller than living pikas.

The fossil record also shows that there were once larger and smaller species of living genera. For instance, a fossil relative of the European rabbit, *Oryctolagus lacosti*, was the size of a hare, while an extinct cottontail, *Sylvilagus leonensis*, was the size of a pygmy rabbit.

Rabbits: The Animal Answer Guide

Chapter 2

Form and Function of Rabbits

What are the largest and smallest living rabbits?

Lagomorphs can be roughly divided into pikas, which are small, rabbits, which are medium-sized, and hares, which are medium-sized to large. Overall lagomorphs exhibit a relatively small range of sizes, from around about 80 grams to about 6 kilograms (2.8 ounces to 15 pounds), but some can weigh as much as 10 kilograms (22 pounds). Compare this with rodents, which range from 6 grams to 66 kilograms (0.21 ounces to 145 pounds). The lagomorphs collectively also fall into a rather odd size class among mammals. The vast majority of mammals, about 80 percent, are small to very small. Most of these are rodents (2,277+ species of 5,416), bats (1,116+ species), and shrews. In fact, some biologists think the optimal size for a mammal is about 100 grams (3.5 ounces), although the modal, or typical, size is about 40 grams (1.4 ounces). For comparison, four slices of Wonder Bread weigh about 100 grams.

Among the lagomorphs, only the smallest of the small pikas are close to the modal size, while larger pikas hit near or exceed the optimal size. The smallest pikas appear to be Thomas's pika (*O. thomasi*) at 45 to 110 grams (1.6 to 3.9 ounces), the Gansu pika (*O. cansus*) at 50 to nearly 100 grams (1.7 to 3.5 ounces), and the Tsing-Ling pika (*O. huangensis*) at 52 to 108 grams (1.8 to 3.8 ounces). Even these diminutive mammals are large compared with most bats, rodents, and shrews—the smallest of which, Remy's pygmy shrew (*Suncus remyi*), tips the scales at a mere 1.8 grams or 0.06 ounces.

Most pikas are somewhat larger, falling in the range between about 100 and 300 grams (3.5 and 10.6 ounces). In only a few pika species do individuals exceed 300 grams (10.6 ounces). The largest species, the alpine

Gansu pikas, native to China, are among the smallest of their family.

Photo © Andrew Smith

pika, weighs from 226 to 360 grams (7.8 to 12.7 ounces), while the Chinese red pika (*O. erythrotis*) weighs from 184 to 352 grams (6.5 to 12.4 ounces). The northern pika, while generally smaller, approaches the size of alpine pikas when they live at the same high latitudes, an illustration of Bergman's rule, which states that animals tend to get larger as temperatures get colder, either due to increasing latitude or increasing elevation.

Pygmy rabbits are the smallest rabbits at from 373 to 458 grams (13.1 to 16.1 ounces), followed closely by volcano rabbits, which range from 387 to 602 grams (13.6 to 21.2 ounces). Among the cottontails, the smallest tapeti ranges from an average of 778 grams to 1.2 kilograms (27.4 ounces to 2.6 pounds), depending on the population, while the largest swamp rabbit reaches 2.7 kilograms (6 pounds). Mexican cottontails (*S. cunicularius*) are also quite large, ranging from 1.8 to 2.3 kilograms (4 to 5 pounds). Among the rest of the rabbits, the smallest are the European rabbit, which average just 1 kilogram (2.2 pounds) in the southwestern population and 2 kilograms (4.4 pounds) in the northeastern one. The Sumatran striped rabbit's weight is estimated at 1.5 kilograms (3.3 pounds). On the large end, the Bunyoro rabbit ranges from 2 to 3 kilograms (4.4 to 6.6 pounds). Domestic rabbit breeds range in size from just under 1 to more than 10 kilograms (2.2 to 22 pounds).

The smallest hares average about 1.5 kilograms (3.3 pounds), and this group includes the snowshoe hare, the Hainan hare (*L. hainanus*), and the Yarkand hare (*L. yarkandensis*) of China. The largest hare is the Alaskan hare, which averages about 4.8 kilograms (10.5 pounds), with the largest reaching 7.2 kilograms (nearly 16 pounds). Body size in hares generally follows Bergmann's rule, with the largest hares living in the far north, the smallest at the equator, and medium-sized species in the temperate zone, although there are clearly exceptions like the snowshoe and Yarkand hares.

Unusual among mammals, in most leporid species females are larger than males, although the differences are fairly small and vary in magnitude depending on whether body weight or linear body measures, such as body length, are compared. This is called "female-biased size dimorphism." The exceptions to this general trend are Alaskan hares and swamp rabbits, in which males and females are about the same size; as noted above, these two species are the largest in their genus.

Biologists think that larger male body size in many species has evolved because of the advantage of large size in winning male-to-male combat over resources, such as receptive females, thus increasing opportunities for mating. Aside from these contest results, females in some species may prefer to mate with larger males because large size may be an indicator of fitness. In other cases, there may be an advantage to males being smaller, if greater agility, for instance, helps them prevail against larger males. This is called the "sexual selection hypothesis." However, female lagomorphs do not usually compete for males, so female size may be determined by what is optimal, under prevailing ecological conditions such as food supply and climate, for producing more young; this is called the "natural selection hypothesis." So, the question is this: have female rabbits and hares been selected to be large, or have males been selected to be small, or both?

A recent detailed study of sexual size dimorphism in cottontails and a few hares, by Candace M. Davis and V. Louise Roth, suggests that there is selection for males to be small. The relationship between female body size and reproductive output in leporids is not clear, but they found that in cottontails, at least, species with larger females do not appear to have greater reproductive success than species with smaller females. Male cottontails engage only minimally in male-to-male combat and achieve mating success by spending time looking for relatively widely dispersed receptive females. Thus the greater agility and mobility that small size provides may be an advantage. Bolstering this hypothesis, the scientists found that there is a tendency for hind foot length to be relatively longer in males than in females. As discussed below (see "Can rabbits run?" later in the chapter), greater hind foot length increases running speed.

There does not appear to be any significant sexual size dimorphism in pikas.

What is the metabolism of a rabbit?

Metabolic rate is speed at which an animal transforms fuel (food), measured in calories, into energy and body tissue. This rate matches the rate of the body's heat production. Among mammals, metabolic rate is closely related to body mass. Overall, smaller mammals relative to their body mass have higher metabolic rates than larger ones, thanks to the geometric relationship between surface area and volume. As an animal gets bigger, its surface area increases (by squaring) but its volume increases much more rapidly (by cubing). Thus a small mammal has a relatively greater surface area to volume ratio than a larger one does. This means that a small mammal loses heat faster than a large one because heat is lost through the relatively larger surface area and less heat is stored in its smaller internal volume.

Other things being equal, it is possible to predict what any particular mammal's metabolic rate is based on its mass alone. But other things are rarely equal, and when measured, a species' metabolic rate may be higher or lower than predicted by its mass alone. Diet, lifestyle, activity level, and climate factors all influence metabolic rate.

As a group, lagomorphs have high metabolic rates for their body size, about similar to those of carnivores, and exceeded only by those of artiodactyls such as deer and cows. This may be because of their use of fermentation to digest their food. It may also be influenced by their relatively short life spans. As a generalization, shorter-lived mammals have higher metabolic rates than long-lived ones. Simply put, they live fast and die young. Further, a high metabolic rate is probably necessary to their ability to escape predation by high-speed running. Species with low metabolic rates are often sluggish and slow moving.

Metabolic rates vary among the lagomorphs. For instance, snowshoe hares have (again, for their body size) higher metabolic rates than desert-living hares, most likely because of the energy it takes for snowshoe hares to maintain body temperature in the cold climate in which they live. For the same reason, pikas have relatively higher metabolic rates than the rest of the lagomorphs. Even individuals may vary their metabolic rate seasonally. For example, the desert cottontail's metabolic rate is lower in summer than winter (see "How do rabbits survive in the desert?" in chapter 5).

An interesting exception among lagomorphs is the arctic hare. Unlike any mammal that lives in a cold environment and unlike all the other lagomorphs, the arctic hare's metabolic rate is actually lower than predicted by its body size, even though these animals maintain about the same body temperature as other lagomorphs. Two features of this species may explain this anomaly. Arctic hares are among the largest lagomorphs, so their surface-to-volume ratio is low and their fur provides extraordinary insulation, relatively more than that of others. Low metabolic rate allows these hares to get by on relatively less food (or less energy-rich food), a plus in the fairly barren habitat in which they live.

Why is rabbit scat round?

Lagomorphs produce two kinds of feces. The cecotrophs, or soft feces, are formed in the cecum, the part of the digestive tract where fine food particles and fluids are shunted for fermentation, while large particles remain in the colon. The soft feces are eaten as they are excreted, so you would not usually see cecotrophs on a walk through the woods. The exception is in pikas, which do not always immediately eat their soft feces, which may be seen near the hay piles they build (see "Do rabbits ever store their food?" in chapter 7).

The hard, dry, roundish pellets you do see derive from the larger, lower-quality fibrous food particles that remain in the colon. Mammals that eat high-fiber diets all tend to have well-shaped, solid feces. As fibrous material passes through the digestive tract, water is also extracted. That the scat is round is probably simply the result of the tubular shape of the lower intestines and the fact that the intestinal muscles contract around the material so that pieces are pinched off, like a string of sausage.

Why do lagomorphs eat their feces, a practice called coprophagy or cecotrophy? Herbivory presents a host of challenges, not least of which is that animals can digest the contents of plant cells but do not produce the enzymes needed to break down the pectins, cellulose, hemicellulose, and sometimes lignin, that make up the plant's cell walls. These cell wall constituents are collectively called fiber. To digest fiber—to extract energy in the form of carbohydrates, animals require the assistance of bacteria and other microbes that reside in their gut. In the gut's anaerobic environment, microbes ferment fiber, consuming it to meet their own energy needs. The by-products of fermentation are called volatile fatty acids, and it is these that herbivores use for energy, absorbing the acids through the tissues of the gut. In addition, the bacteria use nitrogen from plants to produce protein and amino acids that the herbivore can then use.

Mammalian herbivores can be divided into two types, depending on where in the gut this bacterial fermentation takes place. Cows, sheep, deer, camels, kangaroos, some leaf-eating primates, and others are called foregut fermenters. They possess a fermentation chamber, called a forestomach, which lies before the true stomach. Their food rests in this chamber during the fermentation process before passing down through the rest of the digestive tract. Saliva and the mechanical act of chewing starts the breakdown process, which helps speed up microbial fermentation by exposing more surface area for the microbes to attack. Cows and deer, called true ruminants, go a step further and regurgitate large, fibrous particles from the forestomach, rechew it, and swallow the mush—"chew their cud"—for additional mechanical breakdown of fibrous material.

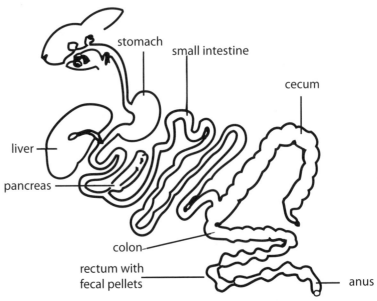

A rabbit's digestive system. Note the size of the cecum compared with the stomach. In humans, the cecum is also known as the appendix.

In contrast, lagomorphs, horses, elephants, and some others are hindgut fermenters. Their microbial fermentation chambers are in the cecum, a pouch connected to the beginning of the large intestines, *after* the true stomach. Saliva and chewing, as in other mammals, start the breakdown process, but only limited digestion occurs in the stomach before indigestible parts of the diet pass into the cecum for fermentation. All mammals have a cecum, but it is greatly enlarged in hindgut fermenters compared with in others. The cecum of lagomorphs is up to 10 times larger than the stomach, and its surface area is enhanced by numerous divisions.

The problem of hindgut versus foregut fermentation is that the nitrogenous protein and amino acid by-products of fermentation, including vitamins, which foregut fermenters absorb in their stomachs, are excreted in hindgut fermenters. Lagomorphs eat their own soft feces so these nutrients can be recaptured when the reingested soft feces are digested in the stomach and small intestine. Rabbits prevented from eating their soft feces starve.

Like the soft feces, hard feces are also regularly reingested to extract as much of the remaining nutrients as possible, but when soft and hard feces are ingested follows a complex pattern. As described by lagomorph specialist Andrew Smith,

> most feeding takes place at night in leporids, and during that time the separation mechanism is activated so that hard feces are formed and fine food particles are shunted to the caecum. In the morning, when animals cease feeding, hard pellets remain in the large intestine. These are excreted and reingested, after which soft feces are formed and reingested until early

Rabbits: The Animal Answer Guide

This series of images, taken with a camera trap, shows a black-tailed jackrabbit in the process of ingesting its soft feces. Photo © Chris Wemmer

afternoon. Then the separation mechanism kicks in again to form hard feces that reach the anus in a few hours and are reingested in the afternoon. Thus, leporids feed on fresh food and ignore feces during the night, and excrete both hard and soft feces during the day and reingest these.

Not much is known about this process in pikas, but there are some differences between European hares and European rabbits. The stomach and cecum are relatively smaller in hares than in rabbits, indigestible fiber passes through the gut of hares faster, and the way hares chew highly fibrous twigs improves the gut's ability to extract carbohydrates.

Some other herbivores may also eat the hard pellets of lagomorphs, which retain some food value because the fiber is so concentrated. Cows, for instance, have been observed to eat European rabbit pellets. American pikas, however, may eat the pellets of marmots (*Marmota* species).

Why are rabbits always sniffing?

Lagomorphs use their sense of smell to communicate with each other (see "Do rabbits talk?" in chapter 4), to detect the presence of predators, and to find food, so it's not surprising that they are always sniffing! Their excellent sense of smell is reflected in the relatively large size of the olfactory lobes in the brain and in the number of olfactory receptor cells in the nose: 100 million in the rabbit compared with our paltry 12 million.

Like other mammals, lagomorphs use two organs to smell, or, more broadly, for olfaction. One involves the chemical receptors on the interior lining of the nostrils, which is what we think of as smell. The other is the

This pygmy rabbit is eating its soft feces. Photo © Jim Witham

vomeronasal, or Jacobsen's, organ, which consists of two tiny openings in the roof of the mouth with ducts that lead to receptors. Chemical messages, usually related to sexual behavior, are sent to the brain via the vomeronasal organ. To draw chemicals into the vomeronasal organ, mammals such as cats perform a behavior called flehmen, or inhaling with the mouth open, to detect scents, most often after sniffing urine. Domestic European rabbits have been observed raising their head and holding it in a horizontal position for several seconds after sniffing urine but do not curl their lips into a grimace as cats do.

Closely related to olfaction, the other chemical sense is taste. The domestic European rabbit has about 17,000 taste buds compared with our 9,000 and a cat's measly 500. Via chemical receptors in the taste buds, rabbits can taste, as we can, sweet, bitter, salt, and sour. In one preference test, rabbits liked the taste of sucrose (sweet) and various salts and disliked that of quinine (bitter) and hydrochloric acid (sour). They also preferred potassium chloride to sodium chloride, the table salt we use. It is interesting to note that the receptors at the front of the tongue, presumably those that make first contact with a food item, are very responsive to bitter and salty and not at all to sweet or sour.

Rabbits can also discriminate between sugars and prefer the sugars that are breakdown products of starch, such as maltose, to the pure sugars, such as sucrose and fructose. Starch is the most abundant carbohydrate in vegetation eaten by rabbits, and their salivary secretions break starch down into maltose during chewing. In contrast, rabbits eat little to no fruit, whose primary carbohydrates are pure sugars. Rats are very similar in this regard

Rabbits: The Animal Answer Guide

Lagomorphs sniff at the round hard pellets of others because the feces convey social information. Photo © John Seidensticker

and have been well studied. Some work suggests that rats have two different types of carbohydrate receptors, rather than a single sweet one. In contrast, humans and other primates show reverse preferences, with sucrose and other pure sugars best liked.

A preference for sweet foods may help rabbits chose nutritious food items but an aversion to bitter ones may help keep them alive to eat another day. Food available to lagomorphs is very often laced with plant secondary compounds, potentially toxic chemicals that plants produce to defend themselves against being eaten by lagomorphs and other herbivores (see "What do rabbits eat?" in chapter 7). Most of these chemicals taste bitter. A taste for salt is generally related to mineral intake.

It's not known whether rabbits have receptors for umami, the savory taste of monosodium glutamate, which were discovered only recently in other mammals. The umami taste detects protein in humans and in other mammals that have been tested, so it is likely that rabbits do have umami receptors. In one experiment, scientists showed that European rabbits preferred high-protein samples to low-protein ones of the same kinds of grass, and taste (or possibly smell or both) seems the most likely way for rabbits to tell one from the other.

Why do rabbits have a "hare lip"?

The split upper lip of lagomorphs, called a philtrum, acts like two pudgy fingers to guide plant parts to their incisors and tongue. It also enables the animals to crop vegetation very close to the ground because the front teeth don't have the lips in the way. Rabbits and hares have slightly different "hare lips." In both groups, the philtrum extends past the teeth to meet the

Form and Function

nostrils, but in rabbits the gums are covered by a layer of skin and in hares the gums are exposed. This gives hares their "buck tooth" look because the upper incisors appear to protrude.

The congenital deformity sometimes pejoratively referred to as "hare lip" in people is more properly called a cleft lip or a cleft palate, depending on which part is involved. It affects about 1 in 850 human babies.

Is it true that rabbit teeth never stop growing?

The teeth of lagomorphs, both their incisors and cheek teeth, do grow continuously throughout their lives. Scientists refer to their teeth as "un-rooted" because the root doesn't close at the base of the tooth but remains open so the teeth can grow. If a lagomorph's teeth didn't continuously grow, the abrasive gritty silica, cellulose, and lignin in their plant food would quickly wear them down to the gums, and the animal would starve to death. In domestic European rabbits, teeth grow on average about 2 to 3 millimeters (0.08 to 0.12 inches) per week, for a total of 105 to 125 millimeters (4 to 5 inches) a year. There is variation in this rate depending on the tooth; lower incisors wear and grow faster than upper incisors. An individual's age, sex, and reproductive status and the pressure applied when the upper teeth meet the lower teeth during chewing, called occlusion, also affect tooth growth and wear. Females, for instance, grow teeth faster than males. In general, the rate of loss equals the rate of growth, but without occlusal wear, the rate of growth would be greater than the rate of loss to abrasives in the diet. Domestic rabbits that lack abrasives in their diet or whose opposing teeth don't meet properly, may suffer from having teeth that grow so long the rabbit can no longer eat. In these cases, their incisors may grow as much as 1 millimeters (0.04 inches) a day!

Lagomorph teeth are arranged so that the sharpness of the cutting surfaces is honed by wearing against the opposing teeth, making the teeth self-sharpening. The softer dentin interior portion of the tooth erodes faster than the harder outer enamel surface, carving the enamel into a sharp-edged ridge. When the enamel ridges of opposing teeth meet, they cut like scissors. The incisors, which are enameled only on the front side, wear to a fine cutting edge, which enables the animal to cleanly slice off bits of vegetation, leaving behind a telltale chisel-shaped cut on stems or twigs.

Rodents also have continuously growing teeth, but in many species with omnivorous diets, including rats and mice, only the incisors do this, while in strictly herbivorous species all teeth grow continuously as in lagomorphs.

Lagomorphs, like most mammals, have only two sets of teeth in their lifetime. Their deciduous teeth, what we call baby teeth, fall out just around

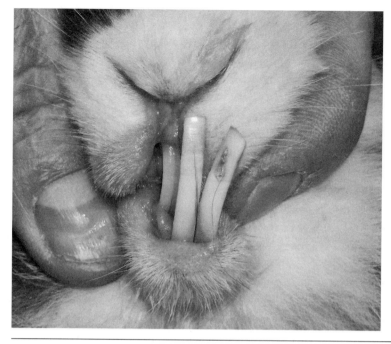

The teeth of lagomorphs grow continuously to make up for erosion caused by a gritty diet. A domestic rabbit fed an improper diet will grow teeth that are so long that it can no longer eat. Photo by Uwe Gille, Wikimedia Commons / CC-BY-SA 3.0

the time of their birth and their permanent teeth are in place by the time young begin eating solid food.

Do rabbits sleep?

All mammals (and most if not all animals) sleep. Sleep seems to be an absolute requirement for mammals, and prolonged sleep deprivation leads to death. The amount of time different mammals sleep, however, varies. The only lagomorph whose sleep has been studied, domestic rabbits, sleep for a total of about 8 hours per 24-hour period. Unlike people, who tend to sleep for about 8 straight hours, rabbits get their 8 hours in multiple short periods of about 25 minutes each.

Most mammals exhibit two kinds of sleep: light sleep and deep sleep, which are differentiated by the brain wave patterns seen in an electroencephalogram. In light sleep, there is very little brain activity and the animal is in quiet repose. During deep sleep—also called REM sleep for the rapid eye movements that occur in this state—intense brain activity is observed. During REM sleep, mammals, including rabbits and people, twitch, jerk, change positions, and so on, but are totally relaxed and more difficult to arouse than they are during light sleep, although once aroused they are more alert than when coming out of light sleep. Domestic rabbits spend about 1 hour in REM sleep per day, or about 11% of their sleep time, compared with about 2 hours of REM sleep in humans, about 25% of sleep time.

Form and Function

What is not known is whether the sleep patterns of domestic rabbits are representative of those of wild lagomorphs. Animals in captivity, which don't have to forage or worry about predators, generally sleep more than animals in the wild. Rabbits newly brought into the novel, stressful lab situation showed no REM sleep for several days.

Among mammals in general, there is a correlation between body size and time spent sleeping, with larger mammals sleeping more than smaller ones. The correlation is particularly strong when only herbivores are considered, although herbivores as a group sleep less than carnivores and omnivores. On these grounds, we might hypothesize that hares will sleep more than rabbits and rabbits more than pikas. There is also a tendency for animals with altricial young to have more REM sleep more than those with precocial young, even as adults, so pikas and rabbits may display more REM sleep than hares.

Outside of the laboratory, scientists cannot usually tell whether an animal is asleep or awake but not active. When not foraging or engaged in social or other behavior, all lagomorphs retreat to cover, either to a burrow or other sheltered spot, and remain relatively still. Rabbits and hares tend to be crepuscular—most active at dawn and dusk with a pronounced midday rest period—but they may be active at any time of the day. Pikas are more strictly diurnal, or preferring to be active in the daylight hours, at least in their surface activity. American pikas, for instance, spend about 30% of daylight hours visible outside of their talus shelters.

Scientists are still not sure why animals sleep. One hypothesis is that sleeping is simply a way to safely pass the time when an animal has satisfied its other needs, although this notion doesn't have a great deal of support. Another proposed function of sleep is that it conserves energy by suppressing activity during a portion of the day. Other ideas with better scientific support include providing time for tissue repair, recharging the supply of neurotransmitters, and learning and the formation of memories.

Can rabbits see color?

Lagomorphs are not color-blind but their perception of color is limited. Two kinds of light-sensitive cells—cones and rods—are found in the retina of lagomorphs, people, and other mammals. Cones detect bright light, and the pigments they possess provide the ability to see color when hit by light of different wavelengths. Rods activate under low light and lack the pigments that produce color vision. In the domestic European rabbit, the only species studied, rods dominate the retina, where rods reach a peak density of about 300,000 per square millimeter in contrast to cones' peak density of about 18,000 per square millimeter. These cones are sensitive to either green or blue light, but not to red, so rabbits are limited to what is called dichromatic (two-color) vision. Most of the retina is dominated by green-sensitive cones, but there is a small area with no green cones and many blue cones. What this may mean for rabbits' color vision is unknown, but the cones would allow rabbits to see over a narrow range of wavelengths during the day. Rabbits don't seem to pay much attention to color though.

The high density of rods means that rabbits can see fairly well in low-light conditions, such as during the dawn and dusk hours. They also have relatively large eyes that may gather more light than smaller ones. However, they may not have the same ability to see in very low light conditions as cats and some other nocturnal mammals do. Rabbits lack the tapetum lucidum behind the retina that these animals possess. Like a mirror, the tapetum lucidum reflects the light that hits it back through the retina to produce a brighter image.

Unlike leporids, pikas are active during the day, have small eyes, and do not seem to rely much on vision, but little is known about how they see. Many mammals that spend much of their time underground, as pikas do, have small eyes and reduced visual abilities.

Other aspects of the leporid visual system are designed to enhance the detection of approaching potential predators. Their large, protruding eyes, set high and well back on the side of the head, give them a visual field of nearly 360°, with a blind spot of only about 10° directly in front of the nose and under the chin. (In comparison, our visual field is about 180°.) This means they can detect movement from all angles, including from above—important when they have to watch for terrestrial as well as aerial predators—without having to move their head. Complementing this, the retina has a horizontal visual streak: an area with a high density of rods that lets the animal focus on all points of the horizon at a time.

The downside of this eye arrangement is that leporids have very poor depth perception, which is the ability to tell how far away things are. This is because they have very little overlap between what each of their eyes sees. When there is extensive overlap, vision is binocular, like ours is. The disparity between what each eyes sees when looking at the same scene is interpreted by the brain to give a three-dimensional view. Binocular vision is very important to predators, such as cats, because it enables the animal to pounce accurately on its prey but is less so to herbivores whose food is stationary. However, rabbits must have some ability to judge distance because they often freeze on the first detection of a predator and run away only when the predator nears.

Leporids appear to have relatively poor visual acuity, which is the ability to distinguish details and shapes of objects. Visual acuity is what is measured in tests of human vision, and glasses or contact lenses are prescribed

to improve visual acuity. So you can imagine that a rabbit has a fuzzy or blurry view of the world, like yours might be if you need glasses but aren't wearing them.

Finally, the lagomorph eye has a nictitating membrane, a transparent or translucent third eyelid that many animals have to protect the eye from injury and help keep it moist. This may account for how rabbits are able to blink just 10 to 12 times an hour. This low blink rate, like all other aspects of rabbit vision, is likely related to the importance of constant alertness to the risk of predation such that even the milliseconds when the blink closes the eyes are dangerous.

Do all rabbits have big ears and short tails?

Compared with other mammals, all lagomorphs have large ears and short tails relative to their body size. Even the smallish ears of pikas are large for their body size compared with those of most rodents. The tails of pikas are not visible, but their hidden tail is actually relatively longer than those of leporids. The ears of cottontails and other rabbits are large but relatively smaller than those of hares. The small tails of cottontails and other rabbits are roundish puffballs, while those of hares are relatively longer and not so wide.

Allen's rule is named for Joel Allen, who in the late nineteenth century stated that mammals from colder climates tend to have relatively smaller ears and tails that do their relatives living in warmer climates. This is because heat is lost through appendages, which increase the surface-to-volume ratio: smaller appendages would help conserve heat in cold climates and dissipate heat in hot ones. This may account for the some of the variation in relative ear and tail size in lagomorphs. The hidden tails and small ears of pikas may be due to the cold climates in which all species live. In rabbits and hares, the ears are known to be used to dissipate heat (see "How do rabbits survive in the desert?" in chapter 5).

Among hares in North America, there is a clear relationship between the length of ears and tails and climate. Relative ear length increases from north to south and east to west. Hares in the desert southwest United States, the antelope jackrabbit (*Lepus alleni*), the black-tailed jackrabbit, and the white-sided jackrabbit (*L. callotis*), have the largest ears, while the snowshoe hares and arctic hares have relatively small ears. Within species there is similar variation, for instance, black-tailed jackrabbits living in warmer areas have relatively larger ears than those in cooler areas of their range. The same holds true for Cape hares (*L. capensis*), in which relative ear length increases with average annual temperature. Among European rabbits in Australia, relative ear length is also greater in warmer areas than cooler ones.

Although there is considerable variation in ear length among cottontails, a clear relationship between the ear length of cottontails and the temperatures in which they live has not been established, although the desert cottontail's ears are relatively longer than those of the eastern cottontail and those of the New England cottontail are shorter. Neither is there a relationship between tail length and temperature.

Ears also have functions that may modify or supplant the effect of temperature on their size. Most obviously, external ears collect sound and, other things being equal, larger ears are better at collecting sounds than smaller ones. Lagomorphs depend in part on sound to detect the approach of predators. Sound travels better through cold air and wet air, so species living in cold or humid climates may be able to get by with smaller ears. Living in the desert, which is both hot and dry, may account for the extremely large ears of some species and populations of hares. Among cottontails, the tympanic bullae, bones that enclose the middle and inner ear, are larger in desert cottontails than in mountain cottontails, whose bullae in turn are larger than those of eastern cottontails. In most mammals, large auditory bullae are correlated with better hearing and are often associated with large ears.

An interesting study of Daurian pikas (*Ochotona daurica*) revealed that populations living at higher and thus colder elevations had relatively larger ears than those living at warmer lower elevations—opposite to the trend predicted by Allen's rule. At the same time, the relative size of the tympanic bullae changes in the opposite direction. The study authors suggest that the stronger winds at higher elevation, which make hearing anything very difficult, as well as the reduction in air pressure, which is known to reduce the hearing ability of people, combine to lead to degeneration of the tympanic bullae. However, hearing remains important for pikas to detect predators. Thus, larger external ears at higher elevations may compensate for the smaller bullae.

Another possible function of lagomorph's long ears relates to locomotion and may account for rabbits have relatively smaller ears than hares. In hares, the long ears act as shock absorbers during high-speed running, helping to reduce jarring movement of the eyes as the front feet hit the ground with force. This may maintain a running hare's visual acuity so it can track the movements of a pursuing predator. A 2010 study by Phillip Stott and his colleagues suggested that this function of the Cape hare's large ears was more important than any role the ears might play in thermoregulation. Rabbits don't achieve the same speeds as hares and so may not need the same relative amount of shock absorption.

Tails in hares and rabbits are used in communication, but are probably not used in maintaining balance during running, as they are, for instance, in kangaroos.

Why do rabbits have whiskers?

All mammals have whiskers. Whiskers, or vibrissae, are long, coarse, stiff hairs that basically extend the sense of touch beyond the surface of the skin. Richly endowed with nerves at the base, vibrissae detect minute variations in air currents caused by objects in the environment. Lagomorphs have 20 to 25 whiskers on each side of the upper lip. Whiskers help mammals navigate in the dark by sensing obstacles to be avoided. They also allow lagomorphs to correctly aim their mouth at a particular part of a plant or a blade of grass and bite it off. Lagomorphs' wide field of vision lets them see almost all around, but they can't see objects in a small area under the mouth. Talus-living pikas have longer whiskers than burrowing pikas do.

Can rabbits run?

All lagomorphs run—running away is a key part of their defenses against predators—and have evolved remarkable adaptations for what scientists call cursorial locomotion. Hares are the most specialized and fastest runners, pikas the least, with cottontails intermediate but closer to hares, and European rabbits and pygmy rabbits, both of which use burrows, closer to pikas.

Most of the skeleton of lagomorphs is designed for speedy running. Their limbs are relatively long, which increases stride length, or the amount

Whiskers, like those on the face of a pygmy rabbit, sense the movement of air and help animals navigate in dim light. Photo © Jim Witham

of ground covered with each step. To further increase stride length, they run on their toes to make the limbs temporarily longer, and they have relatively large feet, especially hind feet. The hind limbs of rabbits and hares are longer than the forelimbs, an important feature for acceleration and jumping because the hind limbs provide more force than the forelimbs. The tibia and fibula (the bones of the hind legs) are fused for greater strength. The spine is flexible, so the animal can flex and arch its back to increase speed. Overall, the skeleton is lightweight.

All of these features are expressed most strongly in hares and least strongly or not at all in pikas. For instance, the clavicle (collarbone) of hares and rabbits is greatly reduced, allowing for freer movement of the front legs. The skulls of hares and rabbits are designed to allow the heavier, face-forward region of the skull to move while the back portion is held still by muscles attached to the neck. This may reduce the jarring effects of limbs hitting the ground with force. As noted above, large ears may similarly act as shock absorbers. Rabbits and hares also have highly muscled hind limbs, particularly in the thighs and hips. This contributes to their power to accelerate and to jump because it increases the force with which they push off the ground and the degree to which the spine flexes.

The gait of running lagomorphs is called a "half-bound," which is used by many small animals for bursts of speed and for moving on uneven terrain. In the half-bound, a rabbit elevates the front of its body, lifts its front legs off the ground, then thrusts forward with both hind feet simultaneously or nearly so. (In contrast, in the gallop of horses, for instance, hind legs push off one at a time.) After push off, the rabbit is briefly in flight—sailing through the air with all feet off the ground. Then, first one front

Form and Function

foot and the other reach the ground, each giving a short thrust that sends the rabbit flying until the hind feet hit the ground again. The forefeet stay on the ground for a shorter amount of time than the hind feet do, because the hind limbs provide most of the propulsive force in the half-bound.

The half-bound hopping of lagomorphs differs from the hopping of kangaroos and some rodents, such as kangaroo rats and jerboas, in that these mammals use only their hind limbs to hop. However, some hares but not rabbits sometimes briefly hop only their hind legs to survey their surroundings—a behavior called "observation leaps"—and the antelope jackrabbit, one of the fastest hares, may take off with a few bipedal hops before switching to the half-bound. Hares also, at very high speeds, may switch from the half-bound to a gallop.

Speed is not the only important aspect of running when it comes to evading predators. Acceleration—how long it takes to reach top speed—and maneuverability—the ability to change direction easily—are also important. A 2006 study comparing greyhounds and European hares demonstrates this. These rivals both reach top speeds of 20 meters (65 feet) per second, equivalent to 72 kilometers (44 miles) per hour, but hares often escape the swift dogs. Hares, it turns out, have much larger hip muscles than greyhounds, which enhance acceleration and maneuverability.

While their gaits are similar, hares, rabbits, and pikas show differences in speed, endurance, and escape strategies. Hares are the fastest, with reported top speeds of 72 kilometers (44 miles) per hour in European hares and antelope jackrabbits, and 64 kilometers (40 miles) per hour in arctic hares and black-tailed jackrabbit. Hares are also greater leapers. Antelope jackrabbits, for instance, can jump over 1.5 meter tall (5 feet) bushes. Cottontails' speeds are reported to range from 24 to 40 kilometers (15 to 25 miles) per hour.

Hares are long-distance runners through open country, while rabbits and pikas sprint for cover that is not very far away. Hares have the endurance to run for long distances—up to about 1.6 kilometers (1 mile) or more if necessary—before exhaustion sets in. At the other extreme, pygmy rabbits can run up to 24 kilometers (15 miles) per hour but only for about 100 meters (325 feet); however this is farther than these rabbits usually venture away from their burrows. American pikas rarely forage more than 6 meters (20 feet) from talus.

Several anatomical features are correlated with hares' greater endurance compared with other lagomorphs. The lungs of jackrabbits, for instance, are relatively larger and more specialized than those of cottontails, which, in turn, are relatively larger and more specialized than those of European rabbits and pikas. The hind limb muscles of hares include more of the fiber types that are resistant to fatigue than those of cottontails, which

While not as speedy as hares, rabbits, like this New England cottontail, are good runners and jumpers.
Photo by David Tibbetts, USFWS

A running hare flies through the air between footfalls that propel its forward movement. Photo by Jeff Kerby, Wikimedia Commons / CC-BY-A 2.0

have more of the fiber types that improve the capacity for quick bursts of movement. This accounts for the different color of the muscles, which are dark in hares but light in rabbits and pikas. Running also builds up heat that must be dissipated, and the large ears of hares dissipate more heat than the smaller ones of rabbits and pikas. Hares also have relatively smaller stomachs and cecums than rabbits and adaptations to more rapidly excrete difficult to digest food particles—this reduces the ballast of the stomach and cecum contents that might slow them down.

Form and Function

Can rabbits swim?

Most mammals can swim, and lagomorphs are not exceptions although not all species have been observed swimming and some live in waterless environments. They generally are reluctant swimmers, taking to the water only when pressed. Hares seem to be stronger swimmers than most rabbits and may swim to escape pursuing predators. Black-tailed and white-tailed jackrabbits are good swimmers and are able to cross rivers, but they usually avoid water. arctic hares swim readily across the small streams that flow through the tundra in the summer and snowshoe hares swim across small lakes and rivers. European hares even dive into rivers to escape predators. It is likely that all cottontails can swim. Eastern cottontails can swim but tend to avoid water. Desert cottontails also swim. Marsh and swamp rabbits are the most water-loving of the lagomorphs. Described as excellent swimmers, these rabbits regularly enter the water to elude predators. European rabbits can swim, and an Amami rabbit was reported to have swum across a stream. Pygmy rabbits sometimes cross streams during dispersal, but their swimming has not been observed. Pikas may be able to swim, but given the short distances they move, probably never have or take the opportunity. Even the best lagomorph swimmers have no special adaptations for efficient swimming. They basically dog paddle, with their eyes and ears, at least, out of the water.

Chapter 3

Rabbit Colors

Why are so many rabbits brown?

The overall coloration of many rabbits and hares appears brownish or grayish, although pikas tend to be reddish to buff colored. But there is a lot of variation in color and color patterns both between and within species, and individual coloration is not uniform over the entire body. The ventral surface of the body (the belly side) is usually lighter than the dorsal surface (the back side).

The most typical mammalian color pattern, characteristic not only of lagomorphs but of most rodents, carnivores, and other mammals is called "agouti," a hue that actually emerges from subtle mixtures of color on individual hairs. An agouti hair is black at the base, yellow in the middle, and black again at the tip, leaving the overall impression of brown to reddish to gray color, depending on the relative amount of two pigments: pheomelanin, which produces red and yellow colors, and eumelanin, which produces dark colors.

Lagomorphs have three types of hairs, and while they all are agouti, they express the pattern differently. In the European rabbit, long guard hairs are black but less dark at the base than at the tips. The hairs of most of the fur are black at the tips, yellow in the middle, and bluish at the base. The short hairs of the underfur (pile) are bluish at the base and tipped in yellow.

Coat coloration generally has at least three functions in animals: concealment from predators (or, conversely, among predators, concealment from prey), communication, and thermoregulation. There are four general ways in which fur coloration is believed to contribute to concealment. One

No lagomorphs are just plain brown. What appears to be a shadow on the back of the neck gives the Indian hare its other common name: the black-naped hare. Photo © Gehan de Silva Wijeyeratne

is background matching, in which an animal's coloration resembles that of its environment. Agouti coloration is so common because it appears to provide good, all-purpose camouflage in a variety of habitats, making it hard for a predator to detect prey, especially when the prey is not moving. Background matching among lagomorphs is revealed in several general trends. They tend to be paler in desert environments where the vegetation and substrate are pale, darker in forests and woodlands, and more gray and reddish in rocky habitats. The rock rabbits of Africa, for example, all have a reddish tone in their fur. The dark brown fur of the volcano rabbit may match the dark soils and basaltic rock of its volcanic habitat.

Black jackrabbits may be the exception that proves the rule. Found only on Espíritu Santo Island in the Gulf of California in Mexico, the black jackrabbit is mostly glossy black and stands out starkly on both green vegetation and bare brown slopes and was described by Howard H. Thomas and Troy L. Bestas as resembling "a short, charred stump." Because this island has no predatory mammals, selection for concealment may have been

Rabbits: The Animal Answer Guide

relaxed, allowing a mutation for melanism, or dark coloring, to persist. Similarly, island forms of the Japanese hare (*Lepus brachyurus*) have darker dorsal pelage (fur) than the mainland ones, and the Amami rabbit is dark, too; in both cases, predatory mammals are absent.

The black jackrabbit does have a few avian predators including, notably, great horned owls (*Bubo virginianus*), one of the few birds of prey able to take adult hares. Many years ago, experiments were done to investigate the ability of owls of different but related species to detect light and dark mice on either light or dark background, given that owls hunt at night in the dark. The owls caught about half as many mice when they matched the background than when they didn't—but only when the substrate was complex, with plenty of places for the mice to hide. On bare substrates, there was no difference. Espiritu Santo Island is fairly barren, with extensive bare brown slopes. In this habitat, a dark hare that looks conspicuous to us may be invisible to a night-hunting owl.

Variable coloring, in which an individual animal's coloration changes as the color of its substrate changes, is also seen in some lagomorphs and is a special case of background matching. Snowshoe hares, for example, change color from brown in the summer to white in the winter, to blend into the snow that blankets their boreal forest habitat in the winter (see "Do fur colors change in different seasons?" below). Also, within species, pelage coloration may vary according to the animal's specific location. Black-tailed jackrabbits, for instance, living on lava fields in an area of Durango, Mexico, have browner fur than those living elsewhere in Durango. The evolution of darker pelage in populations living on lava fields compared with nearby populations living on lighter substrates has been demonstrated in several species of mice.

Another adaptation for concealment is countershading. One thing you might notice is that most species of lagomorphs have lighter fur on their undersides than that on their backs, as do many other mammals. Countershading aids in concealment by reducing the dark shadows cast on the animal's underside by the sun or by making an animal appear flat rather than three-dimensional. The eastern cottontails, for instance, are brown to gray dorsally and white ventrally, while marsh rabbits are dark to reddish brown on top and lighter brown or gray on bottom.

Finally, in disruptive coloration, distinct stripes or spots break up an animal's outline. Among rabbits, only two species, both inhabitants of tropical forests, are distinctly striped: the Sumatran striped rabbit and the Annamite striped rabbit have black stripes running down the back, around the shoulder, and on the face.

Mammals' tendency toward being darker in humid rain forest habitats is known as Gloger's rule, named for Constantin Gloger and first observed in the 1830s. In a detailed comparative analysis of lagomorph coloration, Chantal J. Stoner, Olaf R.P. Bininda-Emonds, and Tim Caro found an association between darker colored species and tropical or subtropical distribution. The Omilteme rabbit (*Sylvilagus insonus*), restricted to humid cloud forest, has reddish dorsal fur with much black, making it appear deep brown. In the Amami rabbit, whose natural habitat was dense subtropical forest, the pelage is dark brown above, reddish brown on the sides, and lighter reddish brown underneath. No hares live in hot, humid rain forests, although there are some mostly very dark to black species, such as the Manchurian hare (*L. mandshuricus*) and the closely related Manchurian black hare (*Lepus melainus*), in which melanistic (dark-colored) individuals are common. Both of these live in cool, dark forests so their darker color may be due to background matching. Another possible reason for the tendency to be darker in humid habitats is that dark fur warms up more quickly in the sun and may thus help keep an animal dry through improved evaporation, and both of these hares live in humid, albeit cool, forests. It is interesting in this light that, among the cottontails, the swamp and marsh rabbits, both denizens of very moist habitats, have dark dorsal pelage, while most of the cottontails are lighter. A recent test of Gloger's rule studied dorsal pelage color of house mice living in 85 different locations in Asia and found a strong association between dark color and more closed humid habitats irrespective of temperature.

Even in species in which the normal color is agouti, melanistic individuals may occur more or less often. Among (nonnative) European rabbits in Tasmania, scientists found that the percentage of black individuals varied with rainfall and elevation. Of the rabbits living above 500 meters (1,640 feet) in elevation where rainfall exceeds 1,250 millimeters (about 50 inches) annually, 20 to 30% are black, while only 2% of rabbits are black on the savanna-like coastal plain, where about 625 millimeters (about 25 inches) of rain falls.

It's possible too that being dark or black may have some other advantages. A 1977 study of European rabbits on Skokolm Island in Great Britain found that black European rabbits lived only a third as long as agouti rabbits, but researchers hypothesized that they persisted in the population because they were less timid and consequently ate more. Better nutrition usually translates into greater reproductive success, which may have compensated for shorter life spans. Since this study, scientists have found in various species of vertebrates that darker or melanistic coat color correlates with enhanced sexual motivation, aggressiveness, reduced responses to stress, and elevated immune responses.

A careful look at the photographs in this book reveals that no lagomorph is just plain brown or gray. Many are washed with black or dark fur, creat-

Although the all-black European rabbit is poorly camouflaged for being out during the day, there may be other benefits associated with melanism (dark coloring) including being able to hide from owls and other nocturnal predators. Photo by T. Voekler

ing a salt-and-pepper look, or have patches of darker or lighter fur than the background color. The ear tips of many hares are fringed in black or the insides white. In many lagomorphs, the tail is conspicuously colored compared with the rest of the body; the undersides of the tails of most cottontails are bright white, for instance. Antelope jackrabbits have bright white rump patches and white-sided jackrabbits have a white rump and white sides.

Cottontails' white tails and the jackrabbits' white patches are most conspicuous when the animals are running from predators, and these sorts of contrasting patterns may work to confuse pursuing predators. For instance, the antelope jackrabbit alternates flashing its left and right rump patches while running in a zig-zag pattern, and the white-sided jackrabbit displays similar moves. The contrasting pattern of black-tailed jackrabbits' ears may work similarly. These jackrabbits display the black side of their tail against their white rumps while running in a zig-zag fashion but do not appear to deliberately flash as the others do. However, scientists recently observed that they do flash their long ears, which on the back side are white with

Rabbit Colors

black tips, by changing from holding both ears erect, to left or right ear erect, to both ears down every 1 to 3 seconds. Similarly, Cape hares flash their black-and-white ears in random order during high-speed running. Called "dazzle coloration," flashing of conspicuous body parts and similar behavior is believed to make it difficult for a pursing predator to determine the speed and trajectory of the moving prey. Dazzle coloration and behavior may be a backup system, employed when camouflage has failed.

For instance, black-tailed jackrabbits raise their ears when they first sense danger, the better to locate the threat through hearing. Then, if the threat is distant, they try to skulk away, running with their body close to the ground and the ears lowered. Or, if the threat is close by, they may first freeze with their head on the ground and ears lowered. Only when this doesn't deter the predator do they burst into speed, display their tails, and start flashing the ears.

Another interesting feature of the coats of rabbits and hares, but not pikas, is a patch of long, dense, fine fur on the upper chest and lower neck, called a "pectoral mane." In all but two species—the volcano rabbit and the Amani rabbit—the pectoral mane fur is brown in contrast to the rest of the ventral pelage. In the volcano and Amani rabbits, the ventral pelage is dark so the pectoral mane doesn't stand out except for its length and texture. The significance of the pectoral mane is obscure, but G. B. Corbet, the scientist who first noticed this feature in a study of many museum specimens, thought its most likely function was to disseminate scent because of its "powderpuff" texture. However, since Corbet published his report in 1982, there has been no further study and only two brief mentions of this feature that we could find.

What causes the fur colors of rabbits?

Processes beginning very early in embryonic development ultimately dictate the fur colors and patterns of the adult animal, processes similar across all vertebrates. In the embryo, neural crest cells running along the spine give rise to melanoblasts—precursors to the melanocytes that produce pigment. The melanoblasts migrate through the body and some enter the epidermis (skin), others the hair follicles, and others the iris of the eye, where they turn into melanocytes. Melanocytes package pigment, or melanin, into melanosomes, which are transferred to the developing hair, skin, or iris cells. Melanosomes are deposited in the inner layers of the hair and give each hair its color or colors, depending on the amount and distribution of pheomelanin and eumelanin, which are defined below.

The major gene locations responsible for hair coloration in mammals have been known for generations. Published in 1930, *The Genetics of Do-*

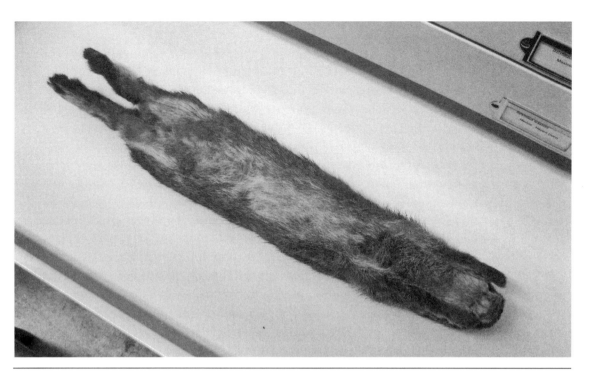

The pectoral mane is visible as a darker patch of fur on this skin of an Omilteme rabbit. Photo © Susan Lumpkin

mestic Rabbits outlined these for rabbits based on mutations observed in domestic animals. Since then, the scientists have identified the actual protein products of most of these genes in mice and discovered many more loci in mice as well, but they are most likely similar or identical in rabbits. The most important gene locations—and those most likely to control pelage color in wild rabbits—are traditionally named agouti, brown, albino, dilute, and extension, abbreviated as *A*, *B*, *C*, *D*, and *E*, respectively. Since the protein products of these gene locations have indentified, scientists name them according to the product.

Agouti (*A*), or agouti signaling protein, produces the normal banded hair of agouti coloration by telling melanocytes to produce pheomelanin, resulting in the middle yellow band. The non-agouti mutation, *a*, produces animals that are uniformly dark and do not have a yellow band in the hair. The non-agouti mutation is recessive, so two *a* alleles are required to produce this color. (An allele is a member of a pair or series of genes that occupy a specific position, or locus, on a given chromosome.) This gene has been sequenced in European rabbits.

Brown (*B*), or tyrosinase-related protein 1, produces eumelanin, dark pigment. A recessive *b* allele produces brown pigment instead of black in an agouti hair. In mice, more than 100 alleles of tyrosine-related protein that cause variation in the production of melanin have been described.

Color (*C*), or tyrosinase, is responsible for the development of pigment at all, and mutations here result in different amounts of melanin production. The recessive *c* gene inhibits the expression of coloring, causing albinism in recessive *cc* homozygotes (individuals with two copies of the same gene). Other mutations at this locus are *cch*, chinchilla, which suppresses color in the middle band of the coat and *ch*, Himalayan, in which only fur at the extremities is black (see "Are there albino rabbits?" below).

Dilution (*D*), or myosin 5a, results in normal pigmentation density of a given color. The recessive mutant *d* affects the intensity of the pigmentation, causing a dilution of the pigment granules. The recessive *dd* dilutes black to blue or yellow to beige.

Extension (*E*), or melanocortin-1 receptor, interacts with the agouti (*A*) gene to cause a shift from eumelanin (dark) to pheomelanin (light). In the presence of the *aa* mutation at agouti, the hair is completely dark. An *e* gene mutation causes increased pheomelanin pigment in the hair, tending to replace the eumelanin, rendering the hair light. Recent molecular studies have uncovered a new mutation of this gene associated with a recessive non-agouti black pelage.

Genetic variations in the agouti gene (*A*) and the melanocortin-1 receptor (*E*) are believed to be the most important source of natural coat color variations by changing the balance between eumelanin and pheomelanin in agouti hairs. Depending on the genetics, switching between pigment types occurs at specific times during hair growth and in particular parts of the body to produce the variety of fur color patterns we see in nature. How these and other gene locations that regulate fur coloration interact to produce the actual colors and patterns exhibited by different rabbit species, and different colors between seasons within a species is not yet known and is just beginning to be understood in mice. In addition to the mutations and interactions between genes noted above, many more have been identified in domestic rabbits that result in the large array of colors and patterns not seen in wild rabbits.

Do fur colors change in different seasons?

Like some other mammals and birds living at high latitudes where the ground is blanketed in snow in the winter, such as long-tailed weasels (*Mustela frenata*), arctic foxes (*Alopex lagopus*), and white-tailed ptarmigan (*Lagopus leucura*), several species of hares living at high altitudes turn white in winter, or at least partly so. As an aside, note that the specific name of the fox and generic name of the ptarmigan is lagopus, from the Greek for hare, *lago*, and foot, *pus*, references to the foxes having fur in the soles of their feet and the ptarmigans having feathered feet.

A snowshoe hare in its white winter coat. Photo from NPS Photos

The fur of the snowshoe hare, once called the varying hare, changes from rusty brown in the summer to white in the winter, although the change is less absolute that this. In the winter, the long overhairs (or guard hairs) are white but only along the parts of these hairs that are longer than fully pigmented underfur. The tips of its ears remain black. Alaskan, arctic, mountain, and Japanese hares also turn from brownish to white in winter, all retaining black on the ears. White-tailed jackrabbits also turn from grayish brown to white. In these hares, unlike the snowshoe hare, the guard hairs are completely white. For all of these species, however, there is significant variation in timing and the completeness of seasonal changes in pelage color, closely tied to the extent and duration of winter snow cover.

White-tailed jackrabbits, for instance, turn mainly white in the snowy northern parts of their range, become mottled with white at the center, and only the sides turn white in the southern parts of their range. Snowshoe hares in the Pacific Northwest, where the ground is seldom covered in snow, do not turn white in winter. Japanese hares turn white in snowy parts of Japan and remain brown elsewhere, and the Irish hare subspecies of the mountain hare (*L. timidus hibernicus*) has patches of white fur among its mostly brown coat in the winter (note, however, the Irish hare may, in fact, be a full species). Among mountain hares, in some areas all fur, including the underfur, turns white while in others the underfur remains dark.

In arctic hares, how long individuals remain white varies with latitude; in the high Arctic, where summer lasts only 6 to 8 weeks, fur on the face may slightly darken but most of the fur never changes from white to dark. During this period of greater vulnerability to predators, when their white

Rabbit Colors

The white-tailed jackrabbit's fur may be mottled in white only in the winter. Photo by Dean Biggins, USFWS

fur stands out against the ground, these hare are warier than they are the rest of the year. Farther south, fur on most of the arctic hare's body changes from white to brownish-gray but the legs, tail, and parts of the ears stay white. Among snowshoe hares in the mountains of Vermont, the duration of the summer pelage varies with altitude: in mid-June, individuals at higher elevations, where snow cover lasts longer, were just losing their white coats while those at lower elevations were in full summer colors.

The timing and extent of pelage color changes appear to be largely genetically controlled and timed to the prevailing average duration of snow cover where the animals live. Snowshoe hares and mountain hares moved between areas continue to change pelage at the same times as they did in their original habitat, so there are not adaptive changes in individuals. However, genetic adaptation may occur very rapidly. In a natural experiment, mountain hares from Norway were introduced to the mostly snow-free Faroe Islands in about 1855. These hares and their descendents at first turned white as they did in Norway. Twenty years later, however, only half turned white; 10 years after that, only a fourth did; and 5 years later, only a few. By the 1930s, none turned white in winter. Most probably, selection exerted by red fox predation drove this change. There is some evidence that hares that change fur color may somehow know when their coat color doesn't match the substrate. John Litvaitis found that mostly white and all-white snowshoe hares occupied areas with more cover, and presumably better protection from predators, when there was no snow on the ground during the spring and fall molting periods.

Rabbits: The Animal Answer Guide

Color is the not the only change to fur that occurs seasonally. Winter pelages are generally thicker, denser, and longer, which increases their ability to insulate against the cold. Fur appears white due to the absence of pigment in most or all of the hairs, and the space occupied by pigment in the summer fur is replaced by air; this too increases insulation. The white winter pelage of snowshoe hares, for instance, provides 27% more insulation than the summer pelage. Even in temperate-zone species without strong seasonal changes in fur color, texture changes from summer to winter. In domestic European rabbits raised for fur, the short, thin summer pelt weighs 50 grams (1.8 ounces), while the longer, denser winter pelt weighs 80 grams (2.8 ounces). The winter coats of American pikas and collared pikas (*Ochotona collaris*) are grayer and nearly twice as long as the summer coat, which lasts only about 2 months.

Other species undergo more subtle seasonal changes in fur color. The winter coats of most pikas are paler and grayer than their summer coats, and the colors are more uniform. For example, in summer, the alpine pika is generally yellowish to ocher to gray with black or dark brown tops and rusty tints on the flanks, while in winter it is gray with a tinge of yellow; the Daurian pika's fur is much lighter in the winter than the straw-gray it is in summer.

The hares that change color seasonally all have two distinct molts—hair shedding—a year: one in the early to late spring to early summer, during which the white winter pelage is gradually lost and replaced by the summer pelage, and the other in the late summer to late fall, when the summer pelage is gradually replaced by the winter one. Lagomorphs without dramatic color changes but with changes in texture may also molt twice a year, such as the European rabbit. American and collared pikas also molt twice a year, as do eastern cottontails. Other species, such as the volcano rabbit, New England cottontails, black-tailed jackrabbits, and the woolly hare (*Lepus oiostolus*), have only one distinct annual molt, in the late summer or fall to winter pelage; summer pelage so gradually replaces the winter one that no obvious molting occurs. The woolly hare of the Tibetan Plateau, by the way, is the only lagomorph with wavy and curly dorsal pelage.

The timing of seasonal changes in pelage color is dictated by photoperiod: the lengthening hours of daylight in the spring trigger the spring molt and the shortening hours of daylight trigger the fall molt. The effect of light on molting, pigmentation, and hair growth is mediated by a structure in the brain called the pineal gland. The pineal gland secretes melatonin in the dark, so the duration of melatonin secretion is proportional to the amount of light reaching the gland via the retina—the longer the period of daylight, the less melatonin secreted and vice versa. Melatonin somehow inhibits melanogenesis—the production of melanin pigment—

Like other species whose winter habitat is cold and snowy, the pygmy rabbit's fur is paler and thicker in winter. Photo © Jim Witham

Woolly hares are the only lagomorph with curly fur. Photo © Andrew Smith

accounting for the loss of pigment that whitens the fur. (This is simplifying quite a bit. A large number of other hormones and neurotransmitters act to influence melanogenesis but a more complete discussion is beyond the scope of this text.)

Melatonin is a very influential hormone, involved in managing circadian rhythms such as the sleep-wake cycle and seasonal rhythms such as in reproductive physiology that are coupled with changing hours of light. It also has a wide array of other physiological effects.

Rabbits: The Animal Answer Guide

What color are a rabbit's eyes?

The eye color of lagomorphs varies from near black or dark brown to yellowish brown and yellow. As with hair, the color is determined by the distribution and pigment content of melanocytes and melanosomes in the iris. The significance of interspecific variation in eye color is not known; the most typical mammalian eye color is brown.

What color are baby rabbits?

Pikas, rabbits, and cottontails are born with little or no fur, although whiskers are present from birth. All hares are born fully furred. Among species for which a description of neonatal and juvenile pelage exists, it is different from that of adults, but not dramatically so.

American pikas are lightly furred and their soft gray juvenile pelage gradually changes to the adult coloration by the end of their first summer. Eastern cottontails at birth are covered in fine fur and have a white spot on the forehead. Hair is sparse in desert cottontails, with no white spot. Swamp rabbit neonates have fur up to about 5 millimeters (0.2 inches) long that is dark on the back, sides, and throat and white on the belly and the chin. Two-day-old Amami rabbits have short brown hair on the body, and the volcano rabbit is born fully furred, with bright yellow fur on the sides, the top of the head, and the extremities. On the back, fur is gray and paler gray on the underside.

At birth, black-tailed jackrabbits are completely covered in fur and already have black ear tips and tail. Unlike the grayish color of the adults, young are brown with accents of yellow and black. A pale gray area surrounds the mouth and nose and there is a patch of white fur on the forehead that mostly disappears by the time the young hare is about 18 weeks old. This richly colored fur of the newborn gradually changes to a paler coat, then to the adult pelage color by the animal's first winter, when it is 6 to 9 months old. Young antelope jackrabbits are slightly darker than adults and also have a white forehead spot that later disappears. Arctic hares are born with mottled, gray-brown fur that provides near-perfect camouflage on the summer tundra. The baby hares then turn white in the fall like their parents do.

Are there albino rabbits?

Albinism is the complete absence of skin and fur color, resulting in an all-white animal with red or pink eyes. In domestic rabbits, albinism is the result of a mutation at the C gene locus, which codes for color saturation, and all albinos are homozygous at this locus, meaning both of the paired alleles have the mutation, noted as *cc*. This combination overrides all other color gene locations. The *C* gene codes for an enzyme called tyrosinase; if this is missing, no pigment is produced. The eyes of albinos are red because the iris and the retina, like the fur and skin, remain unpigmented, making the red blood vessels under the clear retina visible.

While domestic albino rabbits are generally healthy, there are some defects associated with albinism in rabbits, as there are in other mammals, including humans. The visual system is particularly affected in all mammals that have been studied. For instance, albino rabbits have significantly fewer rods (photoreceptor cells in the retina), because melanin synthesis in the retina influences the production of these cells. This means that albinos have a reduced ability to see in low-light situations. The retina also has fewer amacrine and ganglion cells that relay information from the rods and cones to the optic nerve. At the next level of the visual pathway, there are fewer of the optic fibers that go to the same side of the brain (in the mammalian visual system, some optic fibers project to the same side of the brain and others to the opposite side) and they reach the wrong part of the lateral geniculate nucleus, a relay station in the brain. Because of this, regions of the brain that would normally receive a crossed input receive an uncrossed one.

This may, in turn, account for albino rabbits having abnormal nystagmus, or tiny eye movements that keep the retina stimulated and stabilize the image of the visual world on it. Normally, nystagmus consists of alternating eye movement from one direction to the other. However, when the 90 to 180° portion of the visual field of an albino rabbit is stimulated, nystagmus is unstable (but normal in the rest of the visual field). Further, what is known as "opticokinetic nystagmus" occurs in response in looking at moving visual stimuli, such as rotating stripes. In certain parts of the visual field of albino rabbits, instead of the eye following the stimulus, the eye moves away from it. Again, movement is normal in the rest of the visual field. This is possibly due to the ganglion cells in that part of the retina projecting to the wrong side of the albino rabbit's brain. These abnormalities may account for the fact that many albino rabbits exhibit excessive "head-swaying" behavior, sometimes called scanning.

Albino domestic rabbits are the product of human efforts to maintain the trait through selective breeding. Although albinos undoubtedly but very rarely occur in wild lagomorphs, as they do in most mammals, such animals have very poor chances of survival. Lack of camouflage coloration coupled with visual defects means they are easy pickings for predators. With human care and protection, albino rabbits do fine, although they may more readily suffer from exposure to bright light because their unpigmented retinas offer less protection and, lacking skin pigmentation, may more easily be sunburned.

The Himalayan breed of the domestic rabbit exhibits what is called "acromelanic coloration," which is the result of the allele at the *C* locus called *Ch* (for chinchilla). Himalayan rabbits have dark fur on the extremities—nose, ears, and feet—but only if they are raised at 25°C (77°F) or less; at 30°C (86°F) or more, no dark patches develop. In these animals, as in other domestic animals breeds with similar point coloration such as Siamese cats, the expression of the *Ch* allele is temperature-dependent: the tyrosinase it codes stops working at high temperatures. Since the rabbit's core body temperature is greater than 25°C except at the extremities, the rest of its fur is unpigmented. Himalayan rabbits have the same visual system abnormalities as pure albino ones.

In a 2000 experiment that was reported widely in the media and resulted in a great deal of controversy, an albino rabbit was used to create an individual rabbit that fluoresced—lacking other pigmentation, its skin, fur, and eyes glowed green under the proper blue lighting conditions. This was accomplished by inserting a synthetic version of the green fluorescent gene naturally found in a jellyfish into a rabbit zygote. But the goal was not to advance scientific research. Instead, this particular project was the brainchild of Eduardo Kac, a Chicago-based artist, with the assistance of a French zoologist who performed the genetic engineering. Alba, as the fluorescent rabbit was named, was the first known example of what Koc called "transgenic art."

Chapter 4

Rabbit Behavior

Are rabbits social?

The only detailed, long-term studies of the social behavior of lagomorphs are those of a few pika species and the European rabbit. With the exception of the burrowing pika species, lagomorphs tend to be more asocial than not. Most typically, pikas, rabbits, and hares are solitary most of the time, coming together during the breeding season and sometimes forming aggregations at feeding sites.

Even in European rabbits, which live in groups centered on interconnected underground burrows and tunnels called warrens, individuals often react to other members of their group with hostility or avoidance, or, at best, tolerance, provided nobody gets too close. And despite European rabbits' reputation as highly social, their living together may be largely imposed by the limited availability of nest sites, rather than by some innate gregariousness. When nest sites are more evenly distributed, and at low densities, European rabbits may not form social groups at all. Also, there are some distinct disadvantages of group living, as discussed below.

One way to determine how social a species is involves looking at how individuals use space with respect to one another. In lagomorphs, this varies considerably. The American pika, well studied by Andrew Smith and others, is fiercely territorial. Individuals defend their small homes from all others so that there is little overlap in the space they use except among paired males and females with adjacent territories. At the other extreme, in group-living plateau pikas, also well studied by Smith and his colleagues, and in European rabbits, many individuals occupy the same burrows or warrens. In between these extremes are most cottontails and hares, whose home ranges may overlap considerably (depending on density), but except for groups that form at preferred feeding sites, individuals do not simultaneously occupy the same areas within the overlapping home ranges. Hares and rabbits sometimes defend a personal space, which may vary from a less than a meter (39 inches) to several meters.

At the very minimum, males and females must be social to mate, and lagomorphs exhibit a variety of mating systems from monogamy to polygyny (one male and many females) to promiscuity (random mating with no set partners). American pikas form monogamous pairs but the male and female mostly keep their distance except to mate. Male and female arctic hares pair up in the spring and stay together through the breeding season. White-sided jackrabbits also form pairs. In white-tailed, black-tailed, and Tehuantepec jackrabbits, a male's home range overlaps that of several females, and the mating system is polygynous. Eastern cottontail and European hare males tend to form small groups during the breeding season and local males converge on available females in a promiscuous mating system. Within these groups, a dominance hierarchy forms among males, enforced by aggressive behavior. Dominant males mate with most of the females. The system is similar in swamp rabbits, although in this species males protect a territory around breeding females.

Parent-offspring relationships are another aspect of social behavior. Lagomorphs have an absentee maternal care system, in which females visit their young to nurse only once a day and leave them alone the rest of the time (see "Do rabbits care for their young?" in chapter 6). Although mothers and their young in some species do show some affiliative behavior once their young leave the nest, and males often tolerate young—they may even intervene to prevent females from attacking young that may be theirs, as with European rabbits—these relationships end fairly quickly. Adult sometimes aggressively drive away their own and others' youngsters as they approach sexual maturity, or young may basically drift away from their mothers after they are weaned. Plateau pikas are exceptional in this regard. Males provide significant parental care and in the sometimes polyandrous (one female and many males) mating system of this species, young survive better when they have more "fathers."

Dispersal, or permanently moving away from the natal area to live and reproduce elsewhere, is another aspect of social behavior. In many mammals, males disperse farther than females, seeking new home ranges or territories far from those of their mothers. This sex-biased dispersal is thought to minimize the risks of breeding between close relatives, to reduce competition among relatives for mates and resources, or both. Females may also benefit from staying in a familiar place to raise their young. A result of male-biased dispersal is that daughters and sisters often remain close to

Pika species that live in burrows, such as the plateau pika, are more social than those that live in piles of loose rock called talus.

Photo © Andrew Smith

Male plateau pikas provide their young with significant amounts of care. Photo © Andrew Smith

their mothers, called "philopatry," which may enhance cooperation or decrease competition among females because they are related.

There is some evidence of this pattern in European rabbits. A long-term study of European rabbits in a large enclosure in England, led by Diana Bell and Godfrey Hewitt, revealed that males dispersed more often than females before their first breeding season. Genetic analysis then showed that females within a group were more closely related to each other than to females in different groups. For males, in contrast, there was no such pattern.

Rabbits: The Animal Answer Guide

A separate study, part of long-term research on European rabbits in a seminatural environment in Germany, led by Heiko G. Rödel, showed the possible benefits of female philopatry. The scientists found that females that were littermates engaged in very high levels of mutually friendly behavior compared with their interactions with other females. Litter sisters also bred earlier in their first breeding season, perhaps because having a "friend" reduced the stress of living in a social group.

There is evidence of male-biased dispersal in European and mountain hares, snowshoe hares, and in plateau pikas, but not in American pikas, where females may be forced to move farther from their birth place because larger, more aggressive males outcompete them for close-by vacant territories. However, dispersal rates are very low for both sexes and most stay where they were born or nearby, doing their best to avoid their aggressive parents and siblings. Some studies show that dispersal rates are low in mountain hares, too, with fewer than a third moving far enough to leave an adult female's average home range.

Janet Rachlow found that pygmy rabbit females disperse very often and almost as often as males, 80% versus 90%, but female dispersers moved three times farther than male dispersers. Why this should be the case is unclear, but there is some indication that, when female European rabbits disperse, they also move farther than males. Rachlow's research also revealed that tiny pygmy rabbits sometimes dispersed astonishing distances: one female traveled 12 kilometers (7.5 miles) to find a new home! In contrast, dispersing female American pikas move not much more than 100 meters (328 feet).

Feeding aggregations in leporids range from small groups of a handful of individuals to herds as large as 300. These aggregations are temporary and group membership is fluid. The individuals come together to take advantage of a productive feeding site, and relationships among these individuals are tolerant so long as each maintains the proper distance. Dominant individuals may, however, actively displace subordinates from a good patch of food.

Overall, few lagomorphs display what we would call affection toward one another. Harsley Marsden and Nicholas Holler conducted long-term observations of confined eastern cottontails and swamp rabbits. Their monograph, published in 1964, remains the most complete study of cottontail behavior available. They reported no instances of affiliative behavior, such as mutual grooming or resting in body contact, in either species. Minimal affiliative behavior seems to be rule except among the burrowing pikas, although arctic hare pairs have been seen to rest together and sometimes lick and scratch each other peacefully.

The extent to which animals are social depends on the relationship between sociality's costs and benefits. Some potential benefits of social-

Small groups of generally solitary hares form around rich food sources and during the breeding season.

ity in lagomorphs include reduced predation risk due to group vigilance, the ability to harness enough helpers to dig and maintain burrows or warrens, thermoregulation through huddling, and maintaining grazing areas by "managing" the plant community. Another benefit, referred to as the selfish herd phenomenon, is that being in a large group reduces any particular individual's odds of being targeted by a predator. Individuals may also be forced to be social because of the patchy distribution of some essential resource. The costs of sociality include increased risk of predation due to increased conspicuousness, increased competition for food and other resources, and increased disease transmission. For sociality to emerge, the benefits of grouping must outweigh the costs to individuals.

The costs and benefits of sociality are not fixed but may vary with circumstances. For instance, it has been demonstrated that total vigilance increases with group size in European hares. In one study, a hare foraging alone was vigilant about 30% of the time. With each increment in foraging-group size, from 2 to 11, the time during which at least one of the hares was vigilant increased. At the largest group size (8 to 11 individuals) in this experiment, at least one hare was vigilant about 50% of the time. What's more, the time each individual spent in vigilance declined with increasing group size so, consequently, feeding time increased.

This sounds like such a good deal that it seems surprising that these hares don't always feed in groups, but it turns out that it's only a good deal

　　　　Rabbits: The Animal Answer Guide

when food is fairly evenly spaced. If food is in clumps, dominant hares can corner the market and prevent subordinate ones from feeding there—but only up to point. As group size increases, the dominant hares spend a lot of time chasing away competitors from their patch, giving subordinates time to forage in the unprotected patches.

Group size of European rabbits varies but usually consists of several males and several females and their current offspring. In a study of free-ranging rabbits in England by R. M. Lockely, which was the first of its kind, groups consisted of 1 to 8 males and 1 to 12 females. The study of rabbits in Germany mentioned above that observed group size for more than a decade found that groups averaged 2 to 3 males and 3 to 5 females. Usually the males but sometimes females defend their territory from incursions by adults of neighboring groups or strangers, and territorial battles can be fierce at the onset of the breeding season. Within the groups, the males and females formed separate linear dominance hierarchies, also established by intense fighting at the beginning of the breeding season. The males that emerged as dominant win the right to mate with females and maintain this right by directing aggression at subordinates. Dominant females win access to the best nest sites. Once the pecking order is established there is little fighting between females—subordinates tend to avoid dominants and show submission behavior at their approach—but dominant females display far more aggressive than friendly behavior. Dominant males direct friendly behavior almost exclusively to females.

Dominance rank is extremely important to individuals' reproductive success. In the German study, dominant males, or the males in one-male groups, fathered about 90% of all young that survived to adulthood. Dominant females started breeding earlier in the season and had more litters and more young per season and their young grew heavier and more survived to weaning, and, even after weaning, had significantly higher survival rates than those of subordinate females. Males and females that achieved high status in their first breeding season tended to maintain that status in the battles for dominance that marked the onset of their second breeding season. These individuals also enjoyed longer lives than subordinates, although subordinates that achieved dominance in their second year lived longer but not so long as the first group. This longevity effect means that dominant individuals enjoy much greater lifetime reproductive success than subordinate ones. Overall, slightly more than half of females in this study never produced young that survived to adulthood.

These effects may be mediated by stress—subordinate animals are highly stressed by continued aggression, stress leads to reduced resistance to disease, and most of the adult mortality over the winter is due to disease. Stress may also account for subordinate females producing less milk during lactation, leading to their young starving or growing to be smaller than the young of dominant females.

Do rabbits fight? Do rabbits bite?

Yes, they do. For such seemingly meek creatures, lagomorphs sometimes engage in bouts of intense aggression during which fur literally flies. Solitary American pikas resent others intruding on their territories and, when intruders are spotted, give chase. Close encounters may include bumping, kicking with the hind feet, boxing with the forepaws, and biting. In such fights, pikas may lose bits of fur and skin.

Chasing, bumping, kicking, boxing, swiping and scratching with the forepaws, circling, and high leaps over and around the opponent are typical components of rabbit and hare fights. The ears are a favorite target of bites and scratches, so animals may often be seen with torn or scarred ears. Aggression among rabbits and hares erupts primarily in just a few contexts. As in pikas, some fights are over territory, and, depending on the species, may be more common between males and males, females and females, or between members of both sexes. In captive volcano rabbits, for instance, females exhibit territorial aggression toward both males and females, but females fight other females more often and more violently than they do males.

Sometimes violent fights between males erupt during competition over access to a female in estrus, and females not yet quite ready to mate aggressively drive off courting males. Adults of either sex may be aggressive toward subadults, to encourage dispersal or prevent dispersers from settling in their home range. Although several rabbits and hares may forage peacefully in the same area, fights over rich food patches may occur, especially when food is scarce.

Fighting is energetically expensive, and combatants risk injury or death. In confined situations, fights between female European rabbits occasionally end in the death of one of the combatants. Fights to the death have also been reported among wild male snowshoe hares during the breeding season. So, where individuals live in overlapping home ranges or adjacent territories, stable dominance hierarchies, separate for males and females, often emerge. Once an individual has established his or her dominance over others by besting them in a fight, the animals learn their place and subordinates exhibit submissive behavior toward dominants. By conceding space, food, or females to dominants, subordinates avoid possible injury and don't waste time and energy fighting a battle they are likely to lose. Such dominance hierarchies have been observed in European rabbits and in hares and cottontails.

Finally, a female European rabbit sometimes commits infanticide, killing the young of neighboring females while the babies are still in the nest. This may be a way of reducing her own offspring's future competitors. Females guard against this by shows of aggression toward other adult females that approach nest burrows.

How smart are rabbits?

The smart answer is, as smart as they need to be to be a rabbit (or a hare or pika). The measure of intelligence is fraught with difficulty, even when we are dealing only with people. In fact, many scientists doubt the notion that there is a single IQ (intelligence quotient) that reflects much more than the ability to take standardized tests. Instead, some researchers suggest that there are many kinds of intelligence that include linguistic, logical-mathematical, spatial, interpersonal, and kinesthetic. Think about friends who are very good at reading and writing but not in math or are good in math but can't seem to read a map. Which of these people are smarter? It all depends on what you ask them to do. Students of animal behavior confronted a similar problem years ago when tried to test the intelligence of diverse species with the same tests. Cats, for instance, would quickly learn to escape puzzle boxes for a food reward but did worse than pigeons when asked to press a lever for a reward. This didn't mean that pigeons are smarter than cats but only that cats, given their naturally complex predatory behaviors, didn't expect to find prey in this predictably easy way.

People who keep house rabbits are often impressed by their pet's apparent intelligence but what they learn to do is usually somehow related to their natural behavior. Rabbits will learn to use a litter box, for instance, which is comparable to wild rabbits using "latrines" as social signals (see "Do rabbits talk?" below). However, it is very difficult to train them to not chew on furniture as chewing is a natural and essential part of a rabbit's life. With extraordinary patience, rabbits can be taught tricks. Marian Breeland, who studied under pioneering behaviorist B. F. Skinner and was one of the first applied animal behaviorists, trained animals for circus performances and other types of animal shows, as well as for commercials. For one television commercial, made in 1954, she trained a rabbit, known as "Buck Bunny," to drop coins into a model of a bank to sell people on opening a savings account.

Juvenile lagomorphs learn in the wild what's important for their survival probably by observing others: how to respond to alarm calls, to run away from predators, to respond differently to aerial versus terrestrial predators, to interact in socially appropriate ways (such as showing submissive behavior to avoid a fight with a dominant animal), and to navigate around their home ranges (see "Do rabbits play?" later in this chapter).

Lagomorphs, like this brush rabbit, groom their fur with their teeth.

Photo © Chris Wemmer

All that said, biologists remain profoundly interested in comparing intelligence among mammals, in part because of curiosity about how and why human intelligence evolved. One objective measure they look at is brain size, not in absolute terms—although that may be important—but relative to body size, computing a measure called an "encephalization quotient" (EQ). The EQ reflects how much the brain size of a particular species or larger taxonomic group exceeds that expected for its body size, based on mammalian averages. Thus, human beings have a very high EQ with a proportional brain size far larger than that of other primates. Looked at this way, the handful of lagomorph species measured (European rabbit and four hare species) have smaller brains than expected but about the same size as rodents on average and larger than some groups of rodents. However, their EQ is far less than that of carnivores. There is some evidence that predators actually prefer less smart animals as prey, although lagomorphs were not included in this particular study.

Another possible sign that rabbits may not be the brainiest beasts is that the surface of their cerebral cortex is completely smooth, without the convolutions that are present in carnivores, for example, and far more pronounced in humans. These convolutions are believed to increase the area of this part of the brain, which is involved in higher-order mental processing. In contrast, rabbits have relatively large cerebellums, the area of the brain responsible for coordinating movement on a second-to-second basis, not surprising given their high-speed evasive running tactics and their rapid-fire agile aggressive and courtship behaviors.

Rabbits: The Animal Answer Guide

There are many hypotheses about why brain size, which is assumed to reflect something about intelligence, should vary. For example, in some mammalian groups, social species tend to have larger brains than solitary ones, lending credence to the idea that it takes some kind of greater intelligence to deal with the complexities of group living. Another idea is that brain size is related to the habitat complexity: for instance, arboreal species may be more intelligent because they have to maneuver in three-dimensional space, contrasted with terrestrial species, which stay in a flatter area. More generally, large brain size may be related to flexibility and adaptability in the face of any kind of environmental change. Brain size may also be related the size of species' behavioral repertoire. Based only on the ethogram (a list of a species' different behaviors) of a caged domestic rabbit, lagomorphs' relatively small brain size was correlated with a relatively small behavioral repertoire compared with other mammals.

Another hypothesis relates to energetics. Brains are metabolically expensive tissues to support so there may be trade-offs between fueling a large brain versus fueling some other organ. This would be an interesting hypothesis to explore in lagomorphs, given their tight energy budgets. It is also possible that lagomorphs have sacrificed larger brains in the interest of minimizing extra weight that might slow them down. It would be interesting to compare relative brain sizes among a large sample of lagomorphs to test some of these hypotheses. For instance, lagomorphs include species that differ in the extent of their sociality so one might ask whether the social pikas have relatively larger brains than the solitary ones or whether European rabbits have larger brains than other rabbits or hares.

Do rabbits play?

The best word to describe rabbit and hare play may be exuberant, or, less kindly, maniacal. Play behavior includes mad dashes at nothing or at another rabbit, energetic chases of another rabbit, frisky running in circles around the playmate, and stiff-legged jumps straight up in the air like children on pogo sticks. (You can catch a glimpse of lagomorph play by searching for the word "rabbit" on YouTube.)

Play is most common among youngsters. In eastern cottontails and European rabbits, for instance, young are generally playful until they are about 2 months old. But adults also sometimes burst into bouts of play. Arctic hares are notable for their play behavior when they form groups. Adult domestic rabbits engage in lots of play too and seem to enjoy playing with their human companions as well. While play has been described in lagomorphs, it has not been carefully studied.

Scientists have proposed several functions for play behavior, which is seen in the young of many mammals. Play may be a form of physical training, helping the young animals develop the muscles, coordination, and stamina they need as adults to escape from predators. This makes intuitive sense: children who are couch potatoes are not likely to grow up to be Olympic athletes. Play may be about young animals learning and perfecting behaviors that they will use as adults, and lagomorph play behaviors certainly include actions that resemble adult behaviors. In fact, play in young animals looks very much like adult fighting and courtship and, as the animals mature, their play begins to merge into fights. Play-fighting might also be a way for young animals to size one another up: a dominant youngster is likely to turn out to be a dominant adult so play interactions may influence how they interact with one another as adults. Although the idea is untested in lagomorphs, it is possible that this youthful competition affects how young make decisions about whether they will disperse from their natal area.

Why adult lagomorphs might sometimes play is unclear. Perhaps it reduces the tensions that arise when generally solitary adults form groups and is a way of avoiding actual fights with their risk of injury.

Do rabbits talk?

Of course, no lagomorphs really talk as they do in cartoons, but some do use vocalizations to communicate with each other. Most pikas are highly vocal, rabbits less so, and hares tend to be strong, silent types. Hares are rarely heard to make sounds but they do scream or cry if trapped or injured—a call type also reported in rabbits and pikas. It is not known, however, whether these "death screams" serve a communication function, although the eastern cottontail's distress call appears to make those in hearing range alert. Black-tailed jackrabbit females may grunt or growl when driving off an approaching male, and males may squeal after copulation. Antelope jackrabbits growl during courtship and copulation, while European hares and some others grunt during these activities. A captive male snowshoe hare was heard to emit both a whine and a chirplike call when approaching a female nearing estrus. Female arctic hares growl as they approach spots where they nurse their young.

In addition to the distress call, eastern cottontail males and females squeal during copulation and nesting females grunt when an intruder approaches their nest. The swamp rabbit, consistent with its being more social, has a larger vocal repertoire of five calls. Males emit a loud, repetitive squeak when approaching females, and females squeak when approached by males during nest-building. Females quietly chirp when they're being followed by males. Females and possibly males also squeak during copulation. Finally, swamp rabbits give a loud, two-syllable call during a short run after they spot danger; the calling rabbit and others nearby then all adopt an alert posture. Other species that emit such warning calls include the marsh, volcano, Amami, and pygmy rabbits. European rabbits softly grunt during courtship. Amami rabbit mothers call to attract their young as they approach them to nurse, and this species has a unique, pika-like loud call emitted at dusk before they leave their burrows. The nesting young of European rabbits are very vocal, especially in the hour leading up to their daily nursing opportunity, and it's possible that the young of other rabbits also vocalize while hidden in their nests.

The American pika's vocal repertoire consists of eight different calls that the animals make in a variety of situations. Most prominent are the short call and the long call. Emitted by both males and females, the short call acts as an alarm, alerting others in the vicinity to the presence of a predator; it is also a sign of territorial ownership, warning others not to stray into the caller's home. The short call is a shrill, high-pitched squeak or bleat that may be emitted singly or in a series of up to four squeaks. (Listen to a short call at www.bristlecone.org.)

Pikas emitting the short call of alarm are not entirely altruistic. In fact, individuals call less frequently and wait longer to call when they spot a long-tailed weasel (*Mustela frenata*) than when they spot a pine marten (*Martes americanus*). Weasels are a greater threat to pikas than pine martens, so calling less and later may reduce the weasel's ability to detect the pika. The role of the short call in territorial defense was demonstrated in an experiment in which residents were removed from their territories. Nearby pikas took advantage of the vacant area until the experimenters began playing the former resident's short calls there. Male and female pikas on adjacent territories, which are usually a mating pair, sometimes duet, short-calling back and forth to one another. This is believed to enhance their social cohesion without actually making physical contact.

The importance of the short call to defense against predation is suggested by how pikas' behavior changes under windy conditions. High winds muffle the sound of alarm calls, making them less effective. As a result, the time pikas spend inactive increases with wind speed; high-risk behaviors, such as collecting plants for their hay piles (see "Do rabbits ever store their food?" in chapter 7) and running, decreased the most.

The short calls of American pikas in geographically disparate populations differ in note duration, frequency, and number of notes, between one and four, reflecting the fact that there has been little to no gene flow between pika populations separated by long distances. Within populations, short calls are very similar, but there is enough variation for the short call to be used by pikas to identify individuals, so, for instance, a male may recognize his mate by her call. A territory holder may also recognize and tolerate the calls of his or her neighbors, enabling the resident to go on the offense only if a caller is unknown.

The American pika's long call is often called a song and may last for 30 seconds. It sounds a lot like the squeaky sound made when you rapidly wash a window. (Listen at http://encarta.msn.com/media_461517874/ north_american_pika.html.) Males are usually the singers, long-calling primarily during the breeding season, presumably to attract females. However, both sexes may sing in the fall. Other vocalizations include a wail emitted during conflict situations and by unreceptive females when chased by males, a trill emitted by males pursuing females during the breeding season, and a panic call.

The development of vocalizations in American pikas has also been documented. From birth until about 28 days of age, pikas emit a nursing vocalization—a long series of notes that may go on for minutes on end. This vocalization may be functionally similar to purring in kittens, showing contentment and enhancing the bond between mother and young. During this same period, young emit harsh chirps in response to disturbance and a louder distress call in response to greater disturbance. Young begin to utter other calls later, such as the short call, which is first heard at about 12 days of age, and the long call or song, which females begin emitting at about 16 days and males about 32 days, even though as adults males sing more often than females.

There is considerable variation in the vocal repertoires of pikas. The collared pika, which has been well studied by David Hik and his colleagues, has a short call that consists of a single piercing note. It is used like that of American pikas, as an alarm call and in territorial defense. It also varies geographically and indentifies individuals. But this species lacks the long call that is also emitted by most but not all Asian pikas. In fact, two subspecies of Pallas's pika differ in that one has a long call and the other does not. All but a few of the species for which information is available have a version of the short call. Afghan pikas (*Ochotona rufescens*), however, have only a few soft vocalizations and emit a long cry when alarmed. Turkestan red pikas (*O. rutilla*), nicknamed the "silent pikas," have no short alarm call, do not sing, and pairs do not duet; instead, they sound the alarm with a chattering call. The northern pika's long call includes a chattering element and a whistle element. The vocalizations of some pikas can be heard, by humans at least, from considerable distances. The long call of the steppe pika (*O. pusilla*) carries some 2 kilometers (1.2 miles), for example.

Foot drumming or thumping, made famous by the character Thumper in Disney's *Bambi*, is a form of communication displayed by European rabbits, cottontails, some hares, and, perhaps uniquely among the family, the Afghan pika. The signal created by lagomorphs' thumping their large hind feet may be auditory or seismic, that is, transmitted through vibrations in the ground. Foot thumping may communicate alarm or be a threat, or it may signal to a predator that its approach has been detected so pursuit would be a waste of time and energy. It may also be a warning to youngsters not to leave their nests.

While the vocalizations of pikas have garnered the most attention, pikas also use chemical or olfactory communication. In fact, this is the primary means by which lagomorphs communicate, using urine, feces, and secretions from glands on their cheeks or necks, chins, anogenital area (inguinal and anal glands), female nipples, and possibly their eyes. These odorous secretions are rubbed on other individuals, on objects in the environment, and over an individual's own body.

All Asian pikas have a scent gland on their neck, while the two North American species have glands on the cheeks, but they seem to function in generally the same way as home range or territorial markers. Both sexes mark by rubbing the neck or cheek gland on rocks and stones, but males in some species do so more than females, especially during the breeding season. American pikas recognize both individuals and their sex through these scent marks, and presumably other pikas can do the same. This recognition may account for the pika's greater tolerance of its mate's incursions into its territory compared with incursions by unfamiliar animals. The scent marks also likely function in territorial maintenance and defense. In an experiment in which the cheek glands of American pikas were removed, the animals had a harder time keeping intruders out of their territories. In northern pikas, observations revealed that an intruder stimulated a territory owner to scent mark. Only male Gansu pikas cheek rub and only during the breeding season, when they mark areas of home range overlap with other males. During the breeding season, males also rub their anal glands on the ground while defecating. The anal glands of male alpine pikas are larger than those of females but not during the breeding season, suggesting that these marks are more involved in territorial defense than mating behavior. Some Pallas's pikas neck rub on stones and urinate on small piles of their hard feces left on territorial boundaries and areas where there is frequent conflict with neighbors. American pikas have been reported to urinate on hay piles.

In some other pika species, individuals deposit their hard pellets in concentrated spots, usually called "latrines." Whether these latrines play a role in communication or simply reflect a tendency to defecate regularly in the same sites is not known. Among some rabbits, however, latrines do serve a signal function. The hard pellets of European rabbits emerge coated with secretions from the glands on both sides of the anus. Some pellets are defecated anywhere in the home range but others are deposited in latrines. Male European rabbits use latrines more than females do, and the anal glands are larger in males than in females; gland size is also larger in dominant animals and sexually active ones. Females tend to visit latrines most often during the breeding season, which is also when adult males visit most often; males also frequently chin mark, a behavior discussed below, at latrine sites. Females, in contrast, often rest on latrines, which may transfer the odors of dominant males to their body. Precisely what information is conveyed by the anal gland secretions in latrines is not clear; however they do appear, at least, to relate to territory maintenance by males. Territorial swamp rabbits create latrines on top of logs and have larger anal and chin glands than nonterritorial eastern cottontails, which do not create latrines. Nonterritorial hares create latrines at frequently used resting areas, but these do not seem to play a role in communication, and the anal glands of hares are tiny, with no difference in size between males and females, although size increases during the breeding season.

Rubbing the chin gland, or "chinning," is a prominent scent marking behavior of European rabbits. Rabbits chin mark throughout their territory, making it smell "like home," and males chin females during courtship. Like other scent-marking behaviors, the frequency of chinning varies among individuals: males chin more than females and dominant males more than subordinates. Females chin more during estrus than at other times, estrous females sometimes chin courting males, and females prefer the chin gland scent of dominant males to that of subordinates. Females can also identify individuals by the chin scent marks, or at least between a stranger and a familiar animal, and respond to a stranger by increasing their rate of chinning. Chinning thus seems to be involved in both territory maintenance and mating behavior. Dominant male swamp rabbits also chin, in the presence of females, other males, or alone, while eastern cottontails chin only while exploring new areas.

Dominant male eastern cottontails perform a different form of marking: in the presence of a female or one or more subordinate males, they rub the corner of an eye along a twig or the tip of vegetation. Scent-marking with glands around the eyes is known in some rodents and artiodactyls but is not well described in rabbits or hares although both European rabbits and European hares possess Harderian glands—glands found in animals that have a nictitating, or third, eyelid—in the eyes that may be involved in chemical communication.

Inguinal glands, which lie on either side of the vulva or penis, may be involved in sexual attraction, but there is not much evidence for this idea other than the fact that these glands are larger in European hares and in females compared with males, suggesting that the scent or marks from these glands may indicate a female's reproductive condition. However, these glands in female European rabbits do not vary with female reproductive state, and studies of European rabbits suggest they may be involved in recognizing individuals or group members. In experimental situations, females will attack their own young if they have been smeared with a stranger's inguinal gland odors (and, to a lesser extent, a stranger's anal gland odors). In another experiment, males attacked subordinate female members of their group when they were anointed with inguinal gland odors from other males.

Urine is very important in lagomorph chemical communication, especially related to sexual behavior. Both male and female cottontails and hares squirt or spray urine on each other during courtship (see "How do rabbits reproduce?" in chapter 6), and signals in urine may indicate an individual's sex, identity, and age, a male's fertility and dominance status, and a female's readiness to copulate. European rabbits in addition, urine spray adults, juveniles, and young in their group as well as objects in the environmental such as anthills, and at the entrances to burrows.

Relationships between mothers and their young rely on chemical communication as well. This has been very thoroughly studied in domestic European rabbits by Robyn Hudson and her students and colleagues. Lactating females produce a pheromone from glands on their nipples that is essential to their infants' ability to find and attach to a nipple (see "Do rabbits care for their young?" in chapter 6).

Who eats rabbits?

Mentions of the predators that eat rabbits, hares, and pikas are scattered throughout the book, but it is useful to review this and add more information is a single place. For ease of reading, the scientific names of birds and mammals are omitted in this section.

Mammalian carnivores, especially felids, canids, and mustelids (weasels and their relatives), and birds of prey, such as eagles and owls, are the main predators of lagomorphs. Lagomorphs form significant parts of the diets of many of these and minor parts of the diets of many others. The species of both predators and prey, of course, vary geographically and by habitat, as does the relative importance of lagomorphs in a particular predator's diet.

Felids seem to have a particular affinity for lagomorphs. Almost without exception, rabbits or hares are some part of the diets of cats. Even the diminutive, 2 kilogram (4.5 pound) black-footed cat of South Africa occasionally brings down a hare. At the other end of the size spectrum, the huge tiger hunts hares in the Russian Far East. We've discussed elsewhere the critical importance of snowshoe hares to Canada lynx and European rabbits to Iberian lynx. Depending on the study area, hares form anywhere from between 1 to 80% of the diet of Eurasian lynx. Cottontail rabbits contribute to as much as 90% of the bobcat's diet. Volcano rabbits are also important to bobcats where they co-occur, as are jackrabbits. Wildcats, the ancestors of domestic cats, eat European rabbits and hares; in fact, wildcats may specialize in consuming European rabbits and are declining where rabbit numbers have been reduced. Caracals, which range through Africa to central Asia, takes hares and rock rabbits, while sand cats in Africa hunt hares in winter when rodent prey are hibernating.

Pikas are a favorite food of Pallas's cats, small cats native to Central Asia that also eat Tolai hares (*Lepus tolai*). Pikas also formed 16% of the diet of snow leopards in one area of Nepal. Cheetahs in the Serengeti rely on Cape hares when their preferred prey, gazelles, are scarce. Indian hares (*L. nigricollis*) are key prey of leopards in parts of India, and African leopards also take hares. Cottontails and hares also sustain pumas, and cottontails feed jaguars and jaguarundis in parts of their ranges. In southern Chile, nonnative Cape hares now form nearly 80% of the spring diet of Geoffrey's cats.

Canids are also important lagomorph predators. In North America, coyotes are significant predators of snowshoe hares, jackrabbits, cottontails, and pygmy rabbits. Gray foxes in the eastern United States rely on cottontails for up to 64% of their diet in winter and hunt cottontails throughout their range. Red, kit, and swift foxes hunt cottontails and jackrabbits as well. Red foxes also hunt Alaskan and arctic hares, while arctic foxes hunt the young of both species. Red foxes are very important predators of mountain hares in Europe. In Asia, the Tibetan sand fox is dependent on pikas, which are also eaten by corsac foxes. In various parts of their range, gray wolves prey on mountain hares, Alaskan hares, and arctic hares, as well as jackrabbits and snowshoe hares.

Mustelids are well adapted to hunting burrowing lagomorphs as well as those moving in tunnels under the snow, so much so that European ferrets were first domesticated specifically to hunt European rabbits. Across northern Asia, pikas of various species are essential prey to mustelids including ermine, sable, stone martens, steppe polecats, and Altai weasels. In North America, the long-tailed weasel may be the most prominent predator of pygmy rabbits and is also known to hunt volcano rabbits. Weasels, mink, martens, badgers, fishers, and wolverines all take one or more North American lagomorph.

Miscellaneous other mammals also dine on various lagomorphs. Brown bears in Tibet go after pikas; polar bears may kill the odd Alaskan hare. In North America, skunks and raccoons and their ringtail relatives eat cotton-

Plateau pika (*Ochotona curzoniae*).

Photo © Andrew Smith

Gansu pika (*Ochotona cansus*).

Photo © Andrew Smith

Daurian pika (*Ochotona daurica*).

Photo © Andrew Smith

American pika (*Ochotona princeps*).

Photo © Andrew Smith

Glover's pika (*Ochotona gloveri*).

Photo © Andrew Smith

Pygmy rabbit (*Brachylagus idahoensis*). Photo © Jim Witham

Sumatran striped rabbit (*Nesolagus netscheri*). © Wildlife Conservation Society

European rabbit (*Oryctolagus cuniculus*). Photo © Paolo Lombardo

Amami rabbit (*Pentalagus furnessi*).
Photo © Hiromitsu Katsu

It is increasingly difficult to find even the scat of the endangered bristly rabbit (*Caprolagus hispidus*) in its native tall grass habitat in **Nepal and India.** Photos © John Seidensticker

Lower Keys marsh rabbit (*Sylvilagus palustris hefneri*). Photo © Jason Schmidt

Marsh rabbit (*Sylvilagus palustris*).

Photo © BirdPhotos.com

Tres Marías cottontail (*Sylvilagus graysoni*). Photo © Susan Lumpkin

Desert cottontail (*Sylvilagus audubonii*). Photo by H. Cheng, Wikimedia Commons / CC-BY-SA 3.0

Appalachian cottontail (*Sylvilagus obscurus*). Photo © Art Drauglis

Brush rabbit (*Sylvilagus bachmani*).
Photo by Walter Siegmund (with permission),
Wikimedia Commons / CC-BY-SA 3.0

Mountain cottontail (*Sylvilagus nuttalli*). Photo © Jim Witham

Swamp rabbit (*Sylvilagus aquaticus*).

Photo courtesy of PDPhoto.org

Eastern cottontail (*Sylvilagus floridanus*). Photo by William R. James, USFWS

Red foxes and other canids prey on a variety of rabbits and hares. Photo by John Jarvis, USFWS

tails. An introduced mongoose is one of the Amami rabbit's few predators, and catlike genets hunt Bunyoro rabbits in Africa and European rabbits on the Iberian Peninsula. Arctic ground squirrels and red squirrels eat baby snowshoe hares.

Reptiles are less common lagomorph predators, but the Levantine viper (*Macrovipera lebetina*), a large snake reaching 1.5 meters (5 feet) in length, depends on the Afghan pika for food in parts of Turkmenistan. The Amami rabbit's only natural predator is another 1.5-meter-long snake, *Trimeresurus flavoviridis*. Water moccasins (*Agkistrodon piscivorus*) and diamondback rattlesnakes (*Crotalus adamanteus*) feed on immature marsh rabbits, and other rattlesnakes and gopher snakes may similarly take young cottontails. American alligators (*Alligator mississippiensis*) may also prey on marsh and swamp cottontails in the southeastern United States.

Among the predatory birds, golden eagles, which have a wide Northern Hemisphere distribution, seem to have a particular penchant for lagomorphs: they are known to hunt arctic, Alaskan, mountain, European, and snowshoe hares; all of the jackrabbits; several cottontails; and the European rabbit. The Spanish imperial eagle often is a European rabbit specialist but also eats hares, as does its close relative the imperial eagle, which favors hares across much of its central European to western Asian range. The Eurasian eagle-owl, the largest owl in the world, is also a rabbit specialist in Mediterranean habitats, but it hunts hares in some other parts of its range and pikas on the Siberian steppe. The closely related Cape eagle-owl relies on scrub hares in Kenya and also hunts rock rabbits, and Pharaoh's eagle-owl preys on Cape hares in Egypt. The eagle-owls' North American counterparts, the great horned and snowy owls, are also important predators of lagomorphs, taking, depending on their range, pygmy rabbits, cottontails, jackrabbits, snowshoe hares, and Alaskan hares. Various other species of

Golden and other related large eagles are major predators of hares and rabbits. Photo by George Gentry, USFWS

eagles, owls, hawks, falcons, buzzards, vultures, and even corvids (crows and relatives) eat the lagomorphs in their ranges. In South America, almost one-quarter of the diet of Andean condors now consists of introduced European rabbits and hares.

On the Siberian steppe, when the numbers of plateau and Daurian pikas are high, they form 73% of the diet of eagle-owls, 62% of the steppe eagle's diet, and roughly 20% of both the saker falcon's and upland buzzard's. In the same region, the most important prey of ermine and sables are the alpine and northern pikas. In Kazakhstan, as much as 60% of the sable's diet consists of alpine pikas and the stone marten relies on Afghan pikas in one part of Pakistan, where they also support the booted eagle, and Royle's pika (*Ochotona roylei*) in another; Royle's pika is also important to ermine and Altai weasels. The Turkestan red pika also feeds ermine.

John Litvaitis and his colleagues used historical records and other data to show how land-use changes since European settlement, and in the

Rabbits: The Animal Answer Guide

last 200 years in particular, have affected the numbers of bobcats in New Hampshire, largely due to the impact of these changes on New England cottontails, a major prey species. New England cottontails are restricted to young forests with a dense understory of vegetation. When European settlers began clearing New Hampshire's forests for agriculture, opening the forest probably created new habitat for cottontails, and records suggest that bobcats were fairly abundant before the early 1800s.

As ever more land was cleared, bobcat numbers fell. Litvaitis's data show that on average 30 bobcats were harvested per year through the 1800s. In the late 1800s, farms were abruptly abandoned and forests started to replace them. Soon after, by 1915, bobcat harvests rose and ranged from 100 to 400 annually until they peaked at 421 in 1959. By this time, however, the "new" forests had largely matured, thus excluding New England cottontails. Bobcat harvests fell rapidly, and by 1989 bobcats were deemed a protected species in New Hampshire. Bobcat numbers remain very low, as do the numbers of the cottontails.

How changes in numbers of predators track changes in lagomorph numbers is discussed more fully in "Are rabbits good for the environment?" in chapter 5.

Chapter 5

Rabbit Ecology

Where do rabbits live?

All lagomorphs have two fundamental requirements for their living space: they need places to shelter and places to eat in close proximity, often but not always in a mosaic of shelter sites interspersed with feeding sites. The areas occupied by lagomorphs—either their home ranges or territories—must contain both of these or the animals simply can't live there. For just a few species, such as the swamp, marsh, and riverine rabbits, water is also essential.

Shelter provides both protection from predators and a way to keep warm or cool. Places to shelter means different things to different species. For talus-living pikas, obviously, it means a talus field and for burrowing pikas, areas of soil appropriate for digging burrows. Good soil for digging burrows is also important to European rabbits, which tend to prefer sandy soil but burrow in many other soil types as well, and to pygmy rabbits, which need deep, loose soil for their underground burrows as well as the shelter of sagebrush for their surface activities. For cottontails, shelter can be found in brush piles, under herbaceous or shrubby vegetation, in hedgerows, in briar thickets, or under fallen trees—most species don't seem to be very fussy about the form of the cover so long as it provides the shelter they need. The African rock rabbits rely on rocky outcrops for their shelter. Hares, like cottontails, use a variety of shelter types, from the hedgerows used by European and mountain hares in agricultural land to boulders in the barren landscapes occupied by arctic hares. Dense brushy cover is key to the habitat selection of snowshoe hares, and they thus prefer young forests with dense understory vegetation. New England cottontails also need the dense

American pikas and many other pika species live among piles of loose rock called talus. Photo © Andrew Smith

understory found in young forests, and the maturation of forests in the northeastern part of the United States is contributing to their decline.

Similarly, with some exceptions—such as the pygmy rabbit, which relies on sagebrush—the plant species composition available in a particular area appears to be less important than that there are enough edible plants of any kind at all available throughout the year. This dietary flexibility accounts in part for the wide distribution of eastern cottontails (and cottontails in general) and some hares, as well as the ability of European rabbits and European hares to successfully colonize the far-flung places to which they have been introduced. John Flux and Renate Angermann been pointed out, for instance, that there is not a single plant species common to New Zealand and Britain but European hares introduced from Britain to New Zealand thrive.

Lagomorphs generally live in relatively stable home ranges whose boundaries encompass sheltering and feeding areas. In general, home range size varies with body size in lagomorphs, at least among the three groups. Pika ranges are measured in tenths of hectares, those of rabbits in a few hectares, generally 1 to 4 (about 2.5 to 10 acres), and those of hares in tens or hundreds of hectares, generally 10 to 300 (25 to 740 acres). Apart from body size, the home ranges of hares are larger because most hares live in open grassland habitats where they use their running ability to elude predators, unlike rabbits, which rely on quick dashes to cover. Within groups, a variety of other factors appear to be more important than does body size alone.

In territorial species, home ranges are often larger than the areas individuals defend, which tend to be focused on shelter. For instance, American pikas defend territories of around 500 square meters (0.12 acres or roughly one-tenth of a football field), but their home ranges, which include areas for foraging away from talus where they shelter, may be twice or more as large. Similarly, the family-defended territories of alpine pikas are smaller than their home ranges. European rabbits may also defend a small core area around burrows but forage further afield.

Once an adult pika, rabbit, or hare settles in a home range or territory, it very rarely leaves to find a new home. In many species, home ranges shrink and expand seasonally. For instance, during the breeding season, the home ranges of male eastern cottontails often expand as they seek out as many females to mate with as possible, while those of females may shrink as they tend to stay closer to their nests. Larger male home ranges during the breeding season, or year-round, are reported for most species with promiscuous or polygamous breeding systems.

An individual's home-range size may also vary with seasonal changes in the availability of food. In some species, this may mean foraging farther afield in the summer or wet season as succulent vegetation is widely distributed while making do on less preferred plants closer to home during the winter or dry season. In others, the opposite may be true—they may need to forage more widely to find enough to eat in the less favorable seasons than the more favorable ones. In general, within and between species, home ranges overall tend to be larger in marginal habitats than prime ones. Mountain hares living in the boreal forests of Sweden and Finland, for instance, have much larger home ranges—averaging about 200 hectares (2 square kilometers or 494 acres) over the year—than those living in the milder habitat of Ireland, where average home ranges are about 20 to 50 hectares (about 50 to 124 acres).

The distribution of shelter sites relative to foraging sites also influences home-range size. Francisco Palomares and his colleagues compared the variation in European rabbit home-range size in three different habitats in a single area in Spain. In Doñana National Park, rabbits live in scrubland habitat, which is covered roughly equally with grass foraging areas and bushes and scrub for shelter; in grassland habitat where bush and scrub cover only about 10% of the area and grass the rest; or in the ecotone, or edge habitat, where scrubland abuts pastures and bush and scrub cover just under a quarter of the area.

Although grassland habit offered the rabbits the greatest amount and quality of food, the home ranges of rabbits were the largest there because of the very limited amount of shelter area. Rabbits congregate around shelter so there is competition for food close by, forcing the rabbits to travel farther to get enough to eat. In contrast, rabbits in the ecotone, where abundant shelter and high-quality foraging areas were in close juxtaposition, had the smallest home ranges. Home ranges in the scrubland, with plenty of shelter but rather poor forage, were intermediate in size.

Rabbits: The Animal Answer Guide

It may seem obvious to say that home range size may be quite different depending on when in the daily cycle it is measured. Night ranges are often larger than day ranges in European rabbits and hares because the animals travel to forage at night but spend the day sheltering in a smaller area. (In pikas, the reverse is true.) In some situations, the behavior of hares is like that of human commuters. Antelope jackrabbits, for instance, may travel 16 kilometers (10 miles) round trip between their daytime shelters and the alfalfa fields that provide rich foraging opportunities.

Where do rabbits sleep?

Lagomorphs sleep wherever they shelter. Depending on the species, that might be in a burrow, in crevices among rocks, or in more or less shallow forms—the name given to a rabbit's nest or lair—under vegetation or in the lee of boulders and rocks. Except for pikas, most lagomorphs are nocturnal or crepuscular. This is likely an adaptation to the activity rhythms of their predators. Sheltering under cover during the day may help reduce predation by birds of prey, such as eagles, that hunt during the day, while moving into more open areas at night may reduce predation by cats and other carnivores that rely on cover to stalk their prey. Predation on many lagomorphs is always high, however, and there is evidence that some species are highly attuned to change in the risks of predation.

L. Scott Mills and his colleagues showed that among snowshoe hares, for instance, the risk of predation, primarily by Canada lynx, increased during full-moon nights in the snowy winter but not in the summer. In response, the hares reduced how far they moved under the full moon in winter but not in summer, to avoid as much as possible encountering a predator that is more successful hunting in the snow on bright nights; in the summer, the lynx's success did not vary with the full moon, so the hares didn't change how far they moved. Further, hares moved more on cloudy full moon nights than clear ones, and, during a period of unusual snow cover in the summer, they reduced movement under the full moon as they did in winter.

Do rabbits migrate?

Biologists define migration as the regular, usually seasonal movement of a population of animals to and from different geographical regions or habitats, most often in response to changes in food availability between the different regions due to either temperature or rainfall differences. Perhaps one of the most famous of mammalian migrations is that of the wildebeest (*Connochaetes taurinus*) in East Africa, where massive herds of these ungulates travel between the Serengeti and the Maasai Mara following the emergence of new grasses. Very often, migratory mammals, like migratory birds, move between breeding grounds and wintering grounds. In North America, barren-land caribou (*Rangifer tarandus*) migrate in huge herds as much as 5,000 kilometers (3,100 miles) a year between their summer tundra calving grounds and winter boreal forest habitats. Woodland caribou may migrate, too, but over much smaller distances of 15 to 80 kilometers (11 to 31 miles) within the boreal forest.

No lagomorphs exhibit such large-scale migratory patterns, but regular shifts in habitat use have been documented in a few species. Some black-tailed jackrabbits in Utah, for instance, appear to shift between traditional summer and winter habitats, with the wintering areas being those with less snow cover and thus with more accessible food. Scientists documented long-distance (for a hare) movements of between 5 and 35 kilometers (about 3 to 22 miles), although usually less than 10 kilometers (6.2 miles), throughout the year, but such movements peaked in early spring and late fall. Arctic hares in Newfoundland were reported to move about 2.5 kilometers (1.5 miles) between summer and fall ranges. When snow is too hard for mountain hares to dig for buried vegetation, they may move 5 to 10 kilometers (about 3 to 6 miles) to lower ground, and snowshoe hares may move up to 8 kilometers (about 5 miles) when food is scarce.

It is likely that lagomorphs do not migrate because movement expends a great deal of energy and because migration exposes animals to predators in unfamiliar habitats. Although even some tiny birds and butterflies migrate vast distances between summer and winter feeding grounds, few small mammals make such treks. Instead, many small mammals hibernate or have evolved ways to stay active and find food even during the coldest, most barren months. Instead of moving long distances to find food, lagomorphs change their diets, from grasses and forbs, such as dandelions and clover, in the summer (or wet season) to woodier twigs during the winter (or dry season). Pikas gather food in summer and store it to tide them over for the winter (see "Do rabbits ever store their food?" in chapter 7).

Which geographic regions have the most species of rabbits?

We don't know all of the factors that control how many species can co-exist in a given place, but some general patterns in species diversity have been identified. In areas of comparable size, for instance, more species live in the tropics, and the number declines as one goes north or south toward the poles. Lagomorphs as a group defy this rule, however. All pikas and a majority of hares and cottontails live in the northern temperate zone and only a relatively few hares and most of the other rabbits inhabit the tropics. Another general rule of thumb is the more stressed the environment, the fewer species will live there; for example, deserts, tundra, and high mountains generally have fewer species than wet tropical forests. Again, lagomorphs seemingly defy this rule; no pikas or hares and only a few species of cottontails and other rabbits live in wet tropical forests. A general rule that lagomorphs do follow is the relationship between species diversity and habitat complexity. Mountainous areas have greater species richness than flatlands, and landscapes disturbed by fires, floods, landslips, and similar events provide a mosaic of habitats that support a greater number of species.

This is also a tricky question because the larger the area you consider, the more species will live there because by expanding the area under consideration, you also increase the chance of including additional habitat types and species whose ranges are restricted. Say you're just looking at the United States. The state of Montana, at 380,849 square kilometers (147,046 square miles) in the western United States, with habitats that include mountains, grasslands, and arid steppe, is home to eight lagomorphs: America pika, pygmy rabbit, desert cottontail, mountain cottontail, eastern cottontail, snowshoe hare, black-tailed jackrabbit, and white-tailed jackrabbit. With less than a tenth of the area, the state of Maryland at 32,133 square kilometers (12,407 square miles), which ranges from the eastern coastal plain to mountains but is much smaller, boasts two types of lagomorphs today: eastern and Appalachian cottontails; the snowshoe hare that once lived at high elevations there has been extirpated. All told, the United States, excluding Alaska, boasts 17 species, of which 10 are cottontails; including Alaska brings the total to 19, adding the collared pika and the Alaskan hare. Throw in all of Canada and you add only one species, the arctic hare. This is because, despite its vast area, Canada is dominated by just two habitat types: the boreal forest, domain of the snowshoe hare whose United States distribution is restricted to high-elevation forest, and the tundra, where arctic hares rule.

Mexico, at 1,972,550 square kilometers (761,605 square miles) and less than a third the size of the contiguous Unites States, boasts 15 lagomorphs. Nine are cottontails, five are hares; the final species is the primitive volcano rabbit. Seven of these species are shared with the United States, but eight are found only in Mexico. Mexico's varied topography and habitats account for this diversity because although species' distributions may broadly overlap, for the most part the particular areas or habitats they occupy do not.

For instance, three species—two cottontails and a hare—are found only on three separate islands, and a fourth, the Tehuantepec jackrabbit, has a tiny distribution in Pacific coastal dunes. Black-tailed jackrabbits and white-sided jackrabbits occur in semidesert and grassland, respectively.

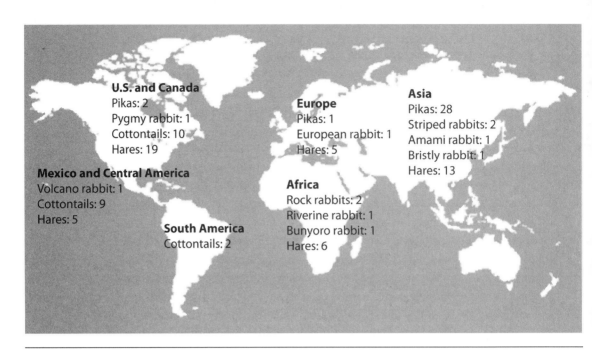

A world map showing the general distribution of lagomorphs. Nonnative ranges of species such the European rabbit and European hare are not shown.

U.S. and Canada
Pikas: 2
Pygmy rabbit: 1
Cottontails: 10
Hares: 19

Mexico and Central America
Volcano rabbit: 1
Cottontails: 9
Hares: 5

South America
Cottontails: 2

Europe
Pikas: 1
European rabbit: 1
Hares: 5

Africa
Rock rabbits: 2
Riverine rabbit: 1
Bunyoro rabbit: 1
Hares: 6

Asia
Pikas: 28
Striped rabbits: 2
Amami rabbit: 1
Bristly rabbit: 1
Hares: 13

Among the other cottontails, the Mexican cottontail lives in temperate forest on the country's western coast, the tapeti is a denizen of tropical rain forest, the Omilteme rabbit inhabits mountain cloud forest, the brush rabbit, as its name suggests, uses brushy habitat in Baja California, and the desert cottontail lives in arid lands. The eastern cottontail is widely distributed but where it overlaps with others geographically it occupies different habitats. For instance, in western Mexico, the eastern cottontail inhabits the high foothills, while the Mexican cottontail occurs along the coastal plain. It is likely, too, that the number of species of cottontails will increase, as the eastern cottontail subspecies, *Sylvilagus floridanus holzneri*, found in the mountains of the southwest United States and north-central Mexico may be raised to species status (see "How many kinds of rabbits are there?" in chapter 1).

On the other side of the world, China can claim 34 lagomorph species; 24 of them are pikas. With 10 species of hares and no rabbits, China, whose area is about equal to that of the United States, has a less diverse assemblage of leporids. There is not much overlap in the geographic ranges of the hares of China, but where there is, they specialize in different habitats. In northwest China, for instance, the desert hare (*Lepus tibetanus*) occupies desert, semidesert, and steppes; the mountain hare lives in subarctic coniferous forests; and the Tolai hare, low-elevation grassland.

Considering all of Asia, there are 45 lagomorphs, again dominated by 28 species of pikas. Only three more species of hare are added to China's total:

Rabbits: The Animal Answer Guide

the Japanese, Indian, and Burmese hares. Asia is also home to four rabbits, with no overlap in their distributions: the bristly, Amami, and two striped rabbits. In total, about half the lagomorphs species live in Asia, but take away the pikas and there are fewer species than in North America, whose eight hares and two rabbits are capped by large radiation of at least 16 cottontails. This is more impressive considering that Asia has about twice the land area of North America.

Six hares, at least four species of rock rabbits, and the riverine and Bunyoro rabbits occupy Africa, plus the European rabbit in the far northwest corner. The only other African species whose distribution extends beyond that continent is the Cape hare. Ethiopia has the most diverse group of hares. Five of the six hare species are found there, and, as elsewhere, each lives in different habitat. The Cape hare prefers open grasslands and semi-deserts. A close relative, the Abyssinian hare (*L. habessinicus*) lives in true desert. The scrub hare (*L. saxatilis*) prefers less open areas such as scrubby grassland and grassy areas within woodlands. The Ethiopian highland hare (*L. starcki*) occupies high-elevation moorlands. And the African savanna hare (*L. microtus*), where it overlaps with the Cape hare, uses scrubbier and higher-elevation habitats. There is no information on the habitat of the Ethiopian hare (*L. fagani*).

Finally, Europe claims the European rabbit and five species of hares, two of which—the mountain hare and the European hare—it shares with parts of Asia. Of the five hares, three are found on the Iberian Peninsula, including the European hare, which may, however, be a relatively recent arrival there. Two are found only on the Iberian Peninsula—the broom hare and the Granada hare, and another, the Corsican hare only in Italy and Sicily (it was introduced to its namesake Corsica). Only South America has as few species, with no hares and a small number of cottontails—currently two named species but most likely several more.

How closely related hares with overlapping geographic ranges maintain their habitat separation is unclear. The widespread European hare and the Corsican hare overlap in central Italy, but the Corsican hare lives at much higher altitudes than the European hare in the area of overlap. However, where the two species do not overlap, the Corsican hare occurs at lower altitudes and the European hare at higher altitudes. Similarly, where they co-occur, the mountain hare lives at higher elevations and the European hare at lower, as do the broom and the European hares in Spain. It could be that as their range expands in Europe, the larger, more adaptable European hares, which also have higher reproductive rates, are displacing the others, pushing them up the mountains into less favorable habitats.

Which rabbits have the largest distributions and which the most restricted?

Biologists use two different measures to describe the size of species' distribution. First is the extent of occurrence, which is defined by essentially drawing a circle around the entire geographic area in which the species is found, or if its distribution is highly disjunct, around each of the one or more areas in which is it found. The second is the area of occupancy, which describes where within the extent of occurrence the animals actually live. Area of occupancy is always smaller than extent of occurrence because it excludes areas of unsuitable habitat. For instance, to cite one example where both are known, the Natal red rock hare (*Pronolagus crassicaudatus*) has an extent of occurrence estimated at more than 20,000 square kilometers (7,722 square miles) and area of occupancy greater than 2,000 square kilometers (772 square miles). The difference can be attributed to the fact that this species relies on habitat—rocky areas and grassy mountain slopes—that is naturally fragmented in coastal southern Africa. Area of occupancy is obviously the better measure to use, especially in the context of evaluating a species' conservation status, but more often than not, it has not been determined.

Let's start with the species with the most restricted distributions because that is the easier question to answer. Some lagomorphs, such as those living on islands and some specialized on rare habitats, have naturally small distributions, but all of these species have also been impacted by habitat loss, fragmentation, and degradation, which has further reduced the size of their ranges.

Although most pikas, which occur on mountain islands—also called sky islands, these are isolated ecosystems occurring at high altitudes—have relatively small distributions, that of the critically endangered silver pika (*Ochotona argentata*) is undoubtedly the smallest. The silver pika occurs only within one small forest where it uses rock outcrops on a single ridgetop area of just 3 square kilometers (1 square mile) in north-central China, and it occupies less than this. The Hainan hare, an island endemic, rivals the silver pika, with no more than 2 square kilometers (0.8 square mile) of optimal habitat remaining on artificially deforested lands currently used as deer ranches.

Several of most range-restricted lagomorphs live in Mexico. The black jackrabbit, endemic to Mexico's Espiritu Santo Island, has an extent of occurrence of about 95 square kilometers (37 square miles)—the area of the entire island—but it probably occupies less than that because habitat has been lost and degraded there. Similarly, the sole island on which the San Jose brush rabbit (*Sylvilagus mansuetus*) lives is 170 square kilometers (65 square miles). The four islands on which the endangered Tres Marías cottontails (*S. graysoni*) live have a total area of about 500 square kilometers (193 square miles), but there has been extensive habitat destruction on these islands and they are described as abundant only on the smallest island with an area of about 60 square kilometers (23 square miles).

Another Mexican species, the endangered Tehuantepec jackrabbit, lives in three separate populations with a total area of occupancy of as little as 67 square kilometers (26 square miles). Like many other range-restricted lagomorphs, this is an endangered species whose range has dramatically declined: its extent of occurrence, now about 520 square kilometers (200 square miles), was once up to 5,000 square kilometers (1,930 square miles). Also endangered, the Omilteme cottontail has an extent of occurrence of less than 500 square kilometers (193 square miles)—if it survives at all—but within that area its habitat is highly fragmented. The endangered volcano rabbit has an estimated extent of occurrence of about 1,840 square kilometers (710 square miles) but occupies fragmented patches on four separate volcanoes totaling about 386 square kilometers (149 square miles).

The distribution of South Africa's critically endangered riverine rabbit, whose extreme habitat specialization means it essentially lives on an island, is not precisely known, with the extent of occurrence estimated at between 100 and 5,000 square kilometers (38 to 1,930 square miles) and area of occupancy at 11 to 500 square kilometers (4 to 193 square miles).

The two Japanese islands on which the Amami rabbit is endemic have total land area of 960 square kilometers (370 square miles), but less than half that area is suitable rain forest habitat for the species thanks to extensive deforestation, and the distribution is fragmented into four separate populations.

The recently described robust cottontail is known from just four mountain ranges in Texas, New Mexico, and northern Mexico, where it lives on mountain islands above 1,500 meters (4,920 feet) in elevation. Its extent of occurrence is about 1,815 square kilometers (700 square miles), but it may occupy only 730 square kilometers (282 square miles). The robust cottontail is considered nearly endangered, and this points to one of the many challenges of lagomorph conservation. Before the robust cottontail was described as a distinct species in 1998, it was considered a subspecies of the widespread eastern cottontail, whose distribution broadly overlaps with that of the robust form. In this status, its conservation would not be a concern, even though it has been extirpated in two, and possibly three, of the ranges it once occupied. Similarly, the broom hare, found only in one mountain range in northern Spain, was once considered conspecific with the widespread European hare. Genetic studies, however, confirmed it is a separate species, and thus, with an extent of occurrence of about 5,000 square kilometers (1,930 square miles) in which it uses a scarce, fragmented heathland habitat above 1,000 meters (3,280 feet), it is now deemed vulnerable to extinction. With so many subspecies and populations of widespread lagomorphs not yet sampled to determine their taxonomic status, it is unknown how many other cryptic species may be in trouble.

Determining the species with the largest distributions is also confounded by taxonomic uncertainties as well as by species whose ranges have been expanded by people. Clearly, if you consider its entire current extent of occurrence, thanks to widespread introductions, the European rabbit wins this competition hands down. Yet, its natural Iberian Peninsula distribution is relatively small. Similarly, the European hare's extent of occurrence was large across Europe and through northern portions of the Middle East, but it has expanded east to parts of Siberia fairly recently and has been introduced in the United Kingdom and southern Scandinavia; outside of Eurasia, it has also been introduced to southern South America, Australia and New Zealand, parts of the United States and Canada, and a few islands as well.

Moreover, some scientists argue based on a variety of genetic evidence that the European hare and the Cape hare, with a large distribution in non-forested parts of southern and northern Africa (but not in the middle) and the Near East, are a single species, in which case its extent of occurrence would be massive. Conversely, others argue that what is now called the Cape hare is one species in southern Africa and another (or several more) in the rest of its range because of the break in the range and no evidence of genetic exchange between them.

On firmer ground, the mountain hare has the largest geographic distribution in Eurasia. Its natural current extent of occurrence stretches across the tundra and taiga (boreal forest) in the northern tier of Eurasia from western Scandinavia to the easternmost tip of Siberia. There are also a few isolated populations of this species in the Alps, in Ireland and Scotland, and in Japan. Next in range is the Tolai hare, which essentially lives in a wide band south of the mountain hare and the European hare's eastern portion, from the eastern shore of the Caspian Sea to the Pacific coast of China. All of the other Eurasian hares have more limited distributions.

In North America, the snowshoe hare's distribution is the largest among hares, stretching across the boreal forest of Alaska, Canada, and the northern tier of the United States from the Atlantic to the Pacific, as well as into New England, through the Rocky Mountain in the west, plus high-elevation sites in California, Utah, and New Mexico. Second place goes to the black-tailed jackrabbit, which is widely distributed in Mexico and through the western half of the United States. Other North America hares have smaller distributions. The eastern cottontail is the most widely distributed of the *Sylvilagus* species and at present its range is expanding.

Finally, among pikas, the northern pika has the largest geographic range of any pika, stretching from the Ural Mountains (which separate Europe and Asia) through northern Asia to the Russian Far East, with a population also found in Japan and some islands is the Bering Sea.

How do rabbits survive in the desert?

In his pioneering study of desert animals, Kurt Schmidt-Nielsen laid out the options of desert dwellers quite simply: "there are three ways in which animals can arrange their life in the hot desert: they can evade the heat, they can passively put up with it, or they can actively combat it by the evaporation of water." Fighting heat through water evaporation, for instance, by sweating, as people do, or panting as dogs do, is dangerous when water is scarce, as it is in the desert. An animal can lose only so much water before it dies, so the challenge is to keep cool without losing too much water.

Several lagomorphs live in deserts where summer daytime temperatures routinely exceed 40°C (104°F) and water is scarce. In fact, desert lagomorphs are not known to drink water as such, instead getting what they need from their food. In the desert, this includes seeking out cacti and other succulent vegetation. Different aspects of adaptations for survival in high desert temperatures have been studied in black-tailed, white-tailed, and antelope jackrabbits and in desert cottontails in the U.S. southwest and in a desert-dwelling population of Cape hares in Israel.

Many desert-adapted animals beat the heat by going underground, spending the hottest part of the desert day resting in burrows, where it is cooler than on the surface; limited activity also reduces heat production. Hares and cottontails are active at night and in the early morning and evening, so their daytime activity is already minimal. For instance, in one study, black-tailed jackrabbits spent most of the daylight hours resting in a shallow form, or lair, under the shade of a shrub and did not move from that position. But on very hot summer days, they became restless in the early afternoon and panting indicated that they were working to keep cool. Usually by this time of day, too, much of the shade they'd found earlier in the day was gone. In this situation, the hares moved to find a form with better shade or moved to burrows, rather exceptional behavior for black-tailed jackrabbits. In some cases, they used desert tortoise (*Gopherus agassizii*) burrows and in others they dug the burrow themselves, gradually excavating a form until it was deep enough to contain the jackrabbit's entire body. Burrow excavation and use appears to be confined to hot summer afternoons when the temperature exceeded 42°C (107.6°F); they were not used in the cold winter or in high wind. The antelope jackrabbit, however, digs out a form that is only deep enough for his haunches to fit into, while the Cape hare uses only a very shallow form.

Such behavioral adaptations are oftentimes not enough, however. At some temperature tipping point, an animal must begin to actively lose heat. Heat is generated both metabolically and, when the ambient temperature exceeds the animal's body temperature, is acquired from the environment. Ways to avoid or reduce heat gain are to reduce metabolic rate and to let body temperature increase. In the jackrabbits and the desert cottontail, metabolic rates are lower in the summer than in the winter, although it may be that the summer rates are "normal" rather than reduced, and it's the winter rates that are elevated. Seasonal comparisons are not available for Cape hares, but comparing the desert-living Cape hares with European hares from the Mediterranean region revealed that the Cape hare's metabolic rate is much lower. The antelope jackrabbit also lets its body temperature rise. Its average body temperature is 37.9°C (100.2°F) at a range of temperatures between 5 and 25°C (41°F to 77°F); above 25°C, it rises with ambient temperature up to about 40.3°C (104.5°F) at ambient temperatures of 39°C (102.2°F). Desert cottontails similarly increase their body temperatures to avoid or reduce heat gain.

Another way to reduce the cost of keeping cool is to shift what is called the zone of thermoneutrality. This is the range of ambient temperatures at which there is no need to expend energy to warm up, at the lower end, or cool down. Both black-tailed and antelope jackrabbits shift their zone of thermoneutrality upward in the summer, but the shift in black-tailed jackrabbits is greater and occurs at both the low and the high end; in contrast, in antelope jackrabbits, it occurs only at the low end, so the thermoneutral zone is narrower in the summer than in the winter. Further, the desert cottontail is similar to the black-tailed jackrabbit in this regard, while the snowshoe hare is similar to the antelope jackrabbit.

Desert lagomorphs also use their remarkable ears to help manage heat gain. For all animals, the larger the body size, the greater the surface area from which heat can be lost. In the relatively small lagomorphs, the surface area of desert-dwellers is increased by their extra-large ears. For instance, the antelope jackrabbit's ears, relatively the largest of all hares, add about 25% to its total body surface area. But all jackrabbits have very large ears. Among black-tailed jackrabbits, which occupy a variety of habitats, the ears of those living in hot temperatures are relatively larger than those in cooler areas. Blood flow to the ears also varies with temperature. When heat loss is possible, that is, when the ambient temperature is lower than body temperature, the blood vessels in the ears are dilated, letting heat dissipate. As ambient temperatures increase, blood vessels contract or narrow to reduce heat gain via conductance. Thermal conductance—which is essentially the

Seemingly exaggerated ears act as radiators to keep desert-dwelling hares from overheating.
Photo by Scott Rheam, USFWS

opposite of insulation—also varies seasonally, with the shorter, lighter fur of the summer pelage making it easier to lose body heat.

At high ambient temperatures, lagomorphs also pant, but they do so only through their nose. For instance, antelope jackrabbits increase their respiration to as high as 700 breaths per minute, compared with 40 per minute at low ambient temperatures. This cools the body through water evaporation but water loss is less than expected based on the animal's body size.

The results of the study comparing desert Cape hares with European hares revealed that Cape hares use water more efficiently, with the value for rates of water turnover about half that expected by their body weight and comparable to the rates for jackrabbits. However, the Cape hare is better able to digest roughage and thus can live on less food than either the European hare or the black-tailed jackrabbit.

Despite these adaptations, desert living is not ideal. For some lagomorphs adapting to the harshness of desert life seems to involve pushing to the limit the adaptations an animal would use to live in less arid habitats that don't have the desert's extreme high temperatures. For example, many jackrabbits die of dehydration and malnutrition when vegetation dries out, although their populations rebound via high rates of reproduction after winter rains. In contrast, Cape hares appear to be better adapted to coping with desert conditions year-round.

How do rabbits survive the winter?

For lagomorphs living in the northern temperate zone, winter is a challenging time. In addition to the energetic costs of maintaining body temperature during long periods of extreme cold temperatures, they must cope with reduced food availability and quality, and, in some cases, the increased energetic costs of locomotion in snow. Lagomorphs display some physiological changes that allow them to cope with winter, but their behavior and the increased insulation of the fur appear to be their primary adaptations for winter survival.

As noted in the previous question, jackrabbits and desert cottontails increase their metabolic rates during the winter and also shift downward their zone of thermoneutrality; the pygmy rabbit may exhibit a similar pattern. In contrast, snowshoe hares decrease metabolic rates. Decreasing metabolic rates reduces the energy necessary to stay alive, which is important during the lean winter months. So why do jackrabbits and desert cottontails take the opposite approach? The increase in energy expenditure due to higher winter metabolic rate is apparently compensated for by a large increase in the insulation provided by their winter fur coats. Fur insulation in the black-tailed jackrabbit is estimated to increase by 42% from summer to winter, while that of the white-tailed jackrabbit increases by about 26% and in snow shoe hares by 27%; pygmy rabbits are also very well insulated generally.

Arctic hares have very low metabolic rates compared with all other lagomorphs, yet their normal body temperatures are similar; they also have extremely low thermal conductance (very high insulation). Not only is the fur longer and thicker in winter, trapping more air (which is what actually insulates), the hollow white hairs of the winter pelage, without the pigment melanin, have more air spaces within the hairs and thus greater insulation.

The other winter adaptations of arctic hares provide a good example of those of other rabbits and hares because they are similar if more extreme. First, arctic hares are among the largest hares anywhere. As body size increases, the surface-to-volume ratio decreases; so large bodies are better able to retain heat. (Overall, within lagomorph species, body size tends to increase with increasing latitudes.) To further reduce heat loss in winter, resting arctic hares expose as little of their surface as possible, especially the less furred extremities. They tuck their tails between their hind legs, their forepaws under the chest, and press their ears down on their back. Thus hunkered down, only their hind feet touch the ground and these are insulated with thick furry pads. Except while foraging, the hares remain essentially immobile in this position to conserve energy.

In winter, the arctic hare's form may be no more than a depression in the snow or in gravel on ridges and slopes, usually facing south on the sheltered side of a rock or boulder, in the sunshine when it is available. They also sometimes dig burrows or dens in the snow. Mountain hares also burrow into snow to find or make a form, where wind speeds are reduced by as much as 90%, as do black-tailed jackrabbits and snowshoe hares.

The cold winter temperatures endured by pygmy rabbits are also offset somewhat by their use of snow burrows, but they do spend more than half of daylight hours on the surface of the snow. In addition to their low thermal conductance, these tiny rabbits rely on the relatively high energy content of the sagebrush that makes up almost the entirety of their winter diets and the shelter offered by the dense sagebrush patches they occupy. These rabbits also forage in tunnels under the snow, as do pikas that live in snow-covered habitats.

Pikas, which survive in very cold high-elevation habitats, have high body temperatures, high metabolic rates, and very low thermal conductance, especially in winter. In American pikas, for instance, fur is nearly twice as long in winter and thermal conductance about half that predicted for a mammal of its size. In addition, pikas employ enhanced "non-shivering thermogenesis" in winter. Shivering is one way to generate body heat, and lagomorphs shiver when temperatures are too cold for them to otherwise maintain their body temperatures. In non-shivering thermogenesis, metabolism of brown adipose tissue—brown fat—generates heat. Non-shivering thermogenesis is regulated by a hormone called leptin, which also helps regulate body weight, body temperature, and metabolic rate. Recent research shows that there has been selection for the genes involved in leptin production in pikas and that there are significant differences in the rate of evolutionary change in leptin genetic patterns between pikas and the European rabbit but not between rabbits and the other species analyzed. Thus adaptive evolution in pika leptin may play an important role in this family's adaptations to cold stress, compared with other lagomorphs.

Do rabbits hibernate?

No, lagomorphs do not hibernate. Instead, they have other adaptations to survive through the winter, as described in the previous answer. A prerequisite for hibernation in mammals in the ability to accumulate the large amounts of fat needed to support metabolism during the winter fast. This requires an energy-rich late summer and fall diet and the ability to consume large amounts of food quickly. By late summer and fall in the north temperate zone, however, the quality of food available to lagomorphs is declining, likely precluding their amassing large stores of fat. However, some species, including the eastern cottontail and the European hare, do have more fat in the winter than in the summer but not enough to enable them to go long periods without food. What's more, a fat lagomorph would have a hard time outrunning its predators.

Do rabbits get sick?

All animals get sick, and lagomorphs are no exception. Lagomorphs acquire internal and external parasites, catch infectious diseases, both viral and bacterial, and are attacked by various kinds of fungi. Both exoparasites (parasites living outside of an animal) and endoparasites (parasites living inside an animal) plague lagomorphs. For example, at least three species of ticks, five species of fleas, and seven species of botflies have been found living on eastern cottontails. The fauna of black-tailed jackrabbits includes various ticks, fleas, lice, and botflies. And a similar assortment of exoparasites is found in lagomorphs everywhere they have been examined. The Amami rabbit, for instance, is host to 13 species of mites and 5 species of ticks. Some exoparasites have specific affinities for lagomorph hosts. *Haemophysalis leporispalustris* is called the rabbit tick because it is primarily found on North and Central American lagomorphs. In other cases, certain exoparasites appear to be unique to a single species. Two mites and a tick are hosted only by Amami rabbits, and two fleas and a mite may be specific to volcano rabbits. The rabbit flea *Spilopsyllus cuniculi* is so closely tied to the European rabbit that the onset of a female flea's reproductive cycle is stimulated by hormones acquired via blood from a pregnant rabbit; the females then are ready to mate and lay eggs timed to be laid on newborn rabbits.

Seven species of sucking lice in the genus *Haemodipsus* also prefer to parasitize lagomorphs. One species is known to live on European hares as well as scrub hares in Africa and has been introduced with the European hares to New Zealand. Another species prefers the European rabbit and it too has been introduced elsewhere, including North America. One native North American species infests several North American hares and at least two species of cottontails. Most recently, another North American native louse in this genus was discovered in pygmy rabbits.

Lagomorphs also host a large array of endoparasites, including cestodes (tapeworms), trematodes (flukes), nematodes (roundworms), and protozoa such as coccidia. The digestive systems of most lagomorphs that have been examined harbor coccidia of one or more species in the genus *Eimeria*; the domestic rabbit may host as many as nine species, one of which, unusually,

Ticks, as seen in the ear of this desert cottontail (*left*), and fleas, like those perched on the edges of this pygmy rabbit's ears (*right*), are two of several ectoparasites that infest lagomorphs. Tick photo by John Mosesso, NBII; flea photo © Jim Witham

lodges in the liver rather than the intestines. In the domestic rabbit, one *Eimeria* causes slight growth retardation in young animals, while another species causes diarrhea and other digestive problems, which are sometimes fatal. A new species of *Eimeria* discovered in captive pygmy rabbits was responsible for several deaths in the captive colony. Scientists comparing the species of *Eimeria* found in pikas revealed that the two species of North America pika host many of the same *Eimeria* species as the northern pika, from which the North American pikas likely evolved. But other Asian pikas host different, diverse species of *Eimeria*. Toxoplasmosis is caused by a protozoan, *Toxoplasma gondii*, which has cats as its final host, but lagomorphs are also commonly, and sometimes fatally, infected.

Species in one group of nematodes are typical internal parasites of lagomorphs, and their evolution and diversification appear to be tied to that of lagomorphs. Species in one genus parasitize only pikas, while its sister group, divided into two genera, parasitizes leporids, suggesting an evolutionary split in this clade at the same time as pikas and leporids split. One of the two genera of these roundworms is found only in the Cape hare and the European rabbit, while the other is Holarctic, with various species occurring in mountain, Tolai, and snowshoe hares, in antelope and black-tailed jackrabbits, and in mountain cottontails.

Parasitic infections are not usually lethal, but they may have debilitating effects. Mountain hares treated with a deworming drug, for instance, had better body condition and may have had increased reproductive success compared with untreated mountain hares. In European rabbits, individuals with more worms weighed less and had less body fat than those with fewer worms, and this may contribute to infected females producing fewer young. High internal parasite burdens were also correlated with the ability of Iberian hares to escape predators.

Lagomorphs also suffer from a variety of viral and bacterial diseases, some of which cause high levels of mortality. For instance, pikas, cottontails, and hares are susceptible to plague, a bacterial disease whose causative agent, *Yersinia pestis*, is transmitted by fleas. Primarily a disease of rodents, lagomorphs get it through contact with infected rodents or from rodents' infected fleas. Tularemia, caused by the bacterium *Francisella tularensis*, is found in hares and rabbits, and epidemics of this bacterial disease occur sporadically. Pseudotuberculosis, from the bacterium *Yersinia pseudotuberulosis*, is also a typical disease of leporids. In North America, cottontails and hares are often infected with Rocky Mountain spotted fever. Various other diseases afflicting lagomorphs include pasteurellosis, brucellosis, staphylococcis, and some herpes viruses. By far the most important diseases of lagomorphs, however, are myxomatosis and rabbit hemorrhagic disease.

Myxomatosis in European rabbits, including domestic forms, is caused by the *Myxoma* virus, which was first identified in South America in 1896 when it killed European rabbits imported there to be used as laboratory animals. Later it was discovered that the *Myxoma* virus was native to tapeti, the widespread South American cottontail, and, in a different strain, to brush rabbits of the western United States. In these two species, the virus causes localized skin tumors, usually at the site of the infection via a flea or mosquito bite, which are fairly benign. *Myxoma* virus is in the family Poxviridae and related to the virus that causes smallpox in people. It belongs to the genus *Leporipoxvirus*, and other members of the genus cause rabbit fibroma, a benign tumor, in eastern cottontails, hare fibroma in European hares, and a fibroma in some African hares, all similarly benign conditions like those that *Myxoma* virus causes in tapeti and brush rabbits.

But in European rabbits the *Myxoma* virus causes systemic infection and a breakdown of the immune system so that secondary infections of pneumonia and other bacterial diseases break out, in much the same way that HIV works to cause AIDS in people. Infection also makes rabbits more susceptible to predation. Symptoms of the disease include tumors and swellings around the head and genitals and then conjunctivitis, blindness, loss of appetite, and fever. Mortality rates may approach 100% in naïve populations of European rabbits, which are uniquely susceptible to the virus's lethal attack. This is due to the virus's ability to circumvent the European rabbit's immune responses, but not those of most other rabbits and hares. Scientists have identified a specific genetic difference between European rabbits and all other lagomorphs examined, including various pikas, cottontails, and hares, that may make it easier for the *Myxoma* virus to invade their cells. The only lagomorph known to share this genetic trait is the riverine rabbit, suggesting that accidental exposure to the virus could wipe out this critically endangered species.

Field-testing the *Myxoma* virus as a means of lethally controlling European rabbit numbers in Australia, where rabbits are pests, began in the late 1930s. In 1950 it escaped into wild there. Within months, the disease spread in southeastern Australia, killing 99%—or more—of the rabbits in its wake. Aided by additional deliberate introductions, by inoculating rabbits with the virus and then releasing them, and natural spread, by 1954 the virus affected all rabbit populations in the country. Inoculation programs then continued, along with the introduction of rabbit fleas, to maintain and enhance the disease's control of rabbit numbers.

In 1952, the virus was released in France by a physician who wanted to eliminate rabbits on his estate. Spreading rapidly through the countryside, within a few years the virus had killed almost all of France's wild rabbits and a large number of its domestic rabbits. Reaching the United Kingdom by 1953, it killed 95% of the rabbits there by 1955. In about the same time frame, it hit rabbits through much of western European including Spain and Portugal, where it devastated the rabbits on their native grounds, killing more than 90%.

The history of the *Myxoma*–European rabbit relationship provides a classic example of what can happen when a virus jumps to a new species and then coevolves with it. After it initially raced through and decimated rabbit populations, two things happened: The small number of rabbits that survived the epidemic developed genetic immunity to it and the virus itself became less virulent, although in some cases virulence increased again in response to rabbits' increasing resistance. As is common in the epidemiology of some diseases, both overly severe and overly mild manifestation of myxomatosis reduced the virus's transmission to new hosts, so there was selection for strains of the virus that produced just the right symptoms for optimal transmission. These less virulent strains kill between 50 and 90% of previously unexposed rabbits, rather than the nearly 100% killed by the most virulent strains they were first exposed to in Australia and Europe. Now, myxomatosis is an endemic disease on both continents and continues to dampen rabbit population growth—a blessing in Australia and a curse on the Iberian Peninsula, where in the 1990s more than a third of young rabbits were still dying of it.

Even as rabbits in Europe were recovering somewhat from the most devastating early epidemics of myxomatosis, a new viral disease came on the scene. First identified in domestic rabbits imported from Europe into China in 1984, rabbit hemorrhagic disease (RHD) is caused by a calicivirus in the genus *Lagovirus*. Transmitted by direct contact and highly contagious, it causes fever, hemorrhaging of the lungs, lesions in the liver, and sick rabbits may bleed from the nose and mouth. Death occurs within 3 days of infection, although rabbits that recover become immune to repeat infections.

Although first seen in China, the virus may have originated in Europe, where scientists have found the related rabbit calicivirus, which doesn't cause disease. RHD reached Europe in 1987 and Spain and Portugal in 1989, where its initial sweep killed 55 to 75% of the peninsula's rabbits. RHD has since been detected in European rabbits, wild and domestic, around the world. In the mid-1990s, it was inadvertently released in Australia, where it killed 90% of some rabbit populations, especially those in dry areas where the virus seems to best thrive there. Only European rabbits appear to be susceptible to this virus.

The *Myxoma* virus and the *Lagovirus* seem to conspire to make life difficult for European rabbits. In Spain and Portugal, for instance, myxomatosis hits in the summer and affects young rabbits, while RHD peaks in the winter and affects adults and subadults.

A closely related calicivirus causes a similar disease in European and mountain hares. Called European brown hare syndrome (EBHS), only these two species are susceptible to the disease. It was first identified in the early 1980s in northern Europe, although it may have been present much earlier, perhaps in a nonpathological form. It has since been found to cause EBHS in hares throughout Europe. Epidemics of EBHS resulted in drops in hare numbers in many areas, but now the disease is endemic and most hares have developed immunity to it, so mortality rates, initially as high as 90%, have fallen sharply.

Are rabbits good for the environment?

Rabbits that have been introduced outside of their natural range, including European rabbits, European hares, and eastern cottontails, can be bad for the environment, often creating trouble for native species of animals and plants. To cite just one example here, nonnative European hares and rabbits favor the growth and spread of nonnative exotic grasses and small plants at the expense of their native counterparts in Chile. The deleterious effects of introduced lagomorphs, especially of the European rabbit, are described more fully in "Are rabbits pests?" in chapter 9. In general, the spread of exotic, invasive species, like the European rabbit, is considered one of the major threats to biodiversity around the world, along with habitat loss and degradation and overconsumption.

Within their natural ranges, however, lagomorphs are part of their environment, so their effects cannot be considered good or bad, just more or less important in their impacts on the ecosystems in which they live. As described earlier, most lagomorphs are critical parts of food chains and webs, providing food for a host of predators.

As herbivores, lagomorphs also have effects on vegetation and thus also on other herbivores. In a study in Denmark, for example, European hares grazing in winter prevented the spread of a saltmarsh shrub, which if left unchecked by the hares would reduce the feeding area available for brant geese (*Branta bernicla*). Some cottontails, the European rabbit, and the European hare are known to act as generalized dispersers of the seeds of some plants, and likely all rabbits do.

Although not native to Britain, the European rabbit in the last 800 years has come to exert significant effects on the landscape and is now considered very important to the maintenance of grasslands in southern England. An interesting finding of a study in Britain is that rabbit grazing increases the colonization of grasses by mycorrhizal fungi, which play important roles in enhancing the growth and vigor of plants and even determine plant biodiversity within grassland ecosystems.

Several lagomorphs play such major roles in ecosystems that they are considered "keystone species": species that have a disproportionately large effect on other species in a community or ecosystem and whose loss in the community would cause significant changes in other species' populations or on ecosystem processes. The snowshoe hare is one such keystone lagomorph. Snowshoe hares are the dominant herbivores of the North American boreal forest, which stretches from the interior of Alaska to the fringes of Newfoundland. The hares are found throughout this vast habitat of mostly low-stature coniferous trees and birch, aspen, and a few other hardy deciduous trees. Canada lynx specialize on hunting hares, and the distributions of the two species match almost perfectly. In fact, predator and prey are so linked that when hare numbers rise and fall dramatically, lynx numbers follow suit.

This ecological phenomenon was well known to boreal forest Amerindians and was first documented by analysis of 300 years of trapping records of the Hudson's Bay Company, which harvested both hares and lynx for their fur. Scientists have been trying to sort out the causes of this cycle for more than 50 years. While huge amounts of research have been conducted, there is not yet a definitive answer, much less a simple one. What's more, the hare and the lynx may star in this system, but they are not the only players on the boreal forest's ecological stage.

The vast literature on this subject was summarized in a 2001 book, *Ecosystem Dynamics of the Boreal Forest*. Edited by renowned ecologist Charles J. Krebs and his colleagues, it reports the results of a massive, 10-year study conducted in the Kluane area of the Yukon. Much of the following comes from this source.

Snowshoe hare density fluctuates in 8- to 11-year cycles, reaching during the peaks densities from five- to twenty-five-fold greater than those during the lows. Each cycle follows a pattern of phases: peak, decline, low, and increase. Changes in lynx densities follow the same pattern, but are delayed about a year and fluctuations are less dramatic, increasing two- to tenfold over the cycle. Two broad hypotheses to explain this have long competed. One argues that changes in food availability for hare drives the system, which ecologists called "bottom-up processes": as hares increase, they eventually reach numbers that outstrip their food supply (the peak); many then starve, reproduction is curtailed, and numbers plummet (the low). This allows their food plants to recover, enabling the hares to increase again. Lynx, being dependent on hares for as much as 80% or more of their food, crash and rebound in turn.

Alternatively, the other idea is that lynx predation on the hares drives the system, a "top-down process." Simply put, when lynx numbers are low, hares flourish; in turn, lynx numbers increase with increased food supply until they begin to kill so many hares their numbers fall. Then the lynx begin to starve and reduce their reproductive output; with fewer lynx, hares begin to flourish again.

Perhaps not surprisingly, both of these processes are in play in the hare-lynx cycle, but predation appears to exert the largest direct effect. In this long-term study, there was no evidence that food was limiting per se; very few hares starved even when their numbers were greatest. However, food quality does appear to have some effect, although it may be indirect. During winter in peak years, hares overgraze their favorite foods; as result, they are in poor body condition as a result of chronic stress. Hares in poor body condition have a harder time avoiding predation, adding to their stress. Stress, in turn, leads to poorer reproduction and sets off the decline phase of the hare cycle. Then, as hares decline, lynx increase, creating further stress on the hares until they hit the low point, whereupon lynx numbers fall, stress is reduced, and hares begin to reproduce more successfully. Thus, lynx not only directly kill more hares during the lynx peak, but they also indirectly dampen hare reproduction through the stress they create on the hares.

But lynx and hares aren't the only players. The boreal forest food web is extremely complex. Snowshoe hares are beset by other predators, including coyotes (*Canis latrans*), whose numbers also rise and fall with those of lynx, as do those of great horned owls, northern harriers (*Circus cyaneus*), and common ravens (*Corvus corax*). Great horned owls don't breed at all in low-hare years. And some prey species similarly track the cycle; muskrats (*Ondatra zibethicus*), for instance, are preyed on by lynx and coyote, so their numbers rise and fall in the same 10-year cycle as the hares. A host of other

Great horned owls of the boreal forest derive a significant part of their diet from snowshoe hares.

Photo by Gary M. Stolz, USFWS

Without snowshoe hares, the boreal forest ecosystem would fall apart.

Photo by Jim Peaco, NPS Photos

species is more or less influenced or influential in this cycle too. It's not surprising that Krebs and his colleagues concluded, "If hares were eliminated, the boreal forest vertebrate community would largely collapse."

Andrew Smith and his colleagues have documented the keystone role of the plateau pika and other pikas in the biodiversity of the Tibetan Plateau ecosystem, where massive poisoning campaigns have eliminated pikas from some areas, providing (unhappily) a way to determine the effects of their loss. In this largely treeless open meadow habitat, many species of birds nest primarily in pika burrows, including Hume's groundpecker (*Pseudopodoces humilis*) and five species of snow finch (*Montifringilla* spp.), and these birds are more abundant in areas with pikas that those without them. The Daurian pika, which inhabits parts of the plateau, provides nesting holes for a snow finch and the Isabelline wheateater (*Oenanthe isabellina*). Two species of lizards also used pika burrows for shelter and breeding. In a census of the birds seen on poisoned and nonpoisoned sites, the snow finches were almost 13 times more abundant and the groundpeckers 20 times more abundant on nonpoisoned sites.

Pikas also affect plant species diversity, primarily by the soil disturbance caused by their burrowing, which loosens and improves the soil. Wildflowers and weeds are more abundant where pikas are present, and some plants grow only on the burrows of either Daurian or Pallas's pika. Pika feces also infuse nutrients into the soil, and plant root weight, plant height, and plant density are all greater in pika areas. Pikas even play a role in ameliorating monsoon flooding on the Tibetan Plateau. Where there are pika burrows, flood waters percolate down into the burrows; without pika burrows, there is more surface runoff that increases downstream flooding.

The little lagomorphs are also the most abundant small herbivores on the Tibetan Plateau, providing food for a large number of predators, which are especially reliant on them in winter, when, because pikas do not hibernate, they may be the only available prey. Steppe polecat (*Mustela eversmanii*) population dynamics are tied to those of pikas, just as their western United States counterparts, black-footed ferrets, were tied to prairie dogs until the decline in prairie dog numbers led to the extinction of these ferrets in the wild (they have now been introduced from ferrets born in captivity). Other small carnivores dependent on pikas on the Tibetan Plateau include two species of weasel, two species of fox, and the Pallas's cat. All of these species disappear when pikas do. Even some large carnivores, such as bears and wolves (*Canis lupus*), rely on a diet of pikas either as their primary food or, in the case of snow leopards, as secondary prey when larger animals are scarce. The diets of brown bears and wolves in some areas consist of 60 and 50% pikas, respectively. The elimination of pikas in some areas has resulted in increased incidence of brown bear depredation on the livestock and crops of Tibetan herders and farmers.

Many of the Tibetan Plateau's large predatory birds are pika connoisseurs as well, and they decline or disappear where pikas are eliminated. Upland buzzards (*Buteo hemilasius*), for instance, are sighted only where pikas are present.

Plateau pikas play a keystone role in the ecology of the Tibetan Plateau.

Photo © Andrew Smith

Pikas are poisoned on the Tibetan Plateau in the belief that they compete with livestock for forage and degrade rangelands through overgrazing. There is no scientific basis for this belief and some evidence that it is simply wrong. As noted above, plants appear to do better where pikas are present than where they are absent. Pikas also prefer areas overgrazed by livestock, so their abundance in overgrazed areas is a symptom, not a cause, of overgrazing. Continued loss of pikas may lead to a collapse in the biodiversity of the Tibetan Plateau.

On the other side of the world, European rabbits have earned new respect as a keystone species in the Mediterranean ecosystems of the Iberian Peninsula, thanks to the work of Miguel Delibes-Mateos and his colleagues. We discuss elsewhere the importance of rabbits as prey in this ecosystem—in brief, at least 29 different predators, including 17 raptors and 9 carnivores eat rabbits either as primary or alternative prey (see "Why are rabbits important?" in chapter 1). With the decline in rabbits due to diseases and habitat alteration, these predators have declined as well. Two, the Iberian lynx and the Spanish imperial eagle, are critically endangered as a result, and the lynx may be extirpated in Portugal. Like the burrows of pikas, the burrows of rabbits provide shelter, refuges, or nesting sites for a plethora of species, including several toads and a newt, snakes and lizards, mice and rats, and some birds. Several carnivores also use rabbit burrows, including Iberian lynx and wolves, which enlarge rabbit warrens to provide temporary shelter for their young. Rabbits host ectoparasites, such as fleas, of at least 22 different genera, and their latrines support several species of Mediterranean dung beetles, some of which appear to have mouth parts specially adapted to crunch dry, fibrous rabbit pellets.

Iberian lynx shelter their young in the burrows of European rabbits, and many other species use rabbit burrows as shelter, nest sites, and refuge from predation. Photo © John Seidensticker

Rabbit pellets are rich in nitrogen and phosphorous, so rabbit latrines, places where dung is concentrated to serve social functions (see "Do rabbits talk?" in chapter 4), create fertile areas between shrubs where plant growth is enhanced, thus contributing to habitat diversity in the semiarid ecosystem. Rabbits also disperse seeds in their dung, and the seeds of at least 72 species of herbs, shrubs, and trees are dispersed in this fashion, helping these species to colonize new areas.

Rabbit grazing has a variety of effects on the plant community. Plants that rabbits prefer, for instance, are less abundant than less-preferred plants, and grazed species are smaller than ungrazed ones. Rabbit disturbance produces patches in which there are fewer, more scattered, and larger shrubs of a particular species, a woody legume under whose canopy a variety of small green plants grow. This habitat of open scrubland with a mosaic of open areas, cover, and the ecotones between them is good rabbit habitat, so rabbits are essentially managing the land for themselves. But other herbivores, including many small mammals, also prefer ecotone habitats so rabbit activity may enhance their abundance, while their greater risk of predation in open areas may benefit predators as well.

In sum, wrote Delibes-Mateo and his colleagues, "Rabbits are so important in the Mediterranean scrubland of southwestern Europe that this ecosystem could be considered 'the rabbit's ecosystem.'" If the continued decline of rabbits there cannot be reversed, the future of this landscape and its flora and fauna is in doubt.

Rabbits: The Animal Answer Guide

Chapter 6

Reproduction and Development of Rabbits

How do rabbits reproduce?

You've probably heard the simile "mad as a March hare" and are familiar with the March Hare character of *Alice in Wonderland*. A similar expression dates to around 1500, "Thanne they begyn to swere and to stare, And be as braynles as a Marshe hare." The allusion is to the mating behavior of European hares, which if not mad, certainly looks madcap. During the breeding season, which extends before and well past March in most areas, the normally secretive hares engage in extended chases, wild leaps, circling, and boxing. Boxing was once thought to be male-on-male aggression, but more often it is a pursued female that punches her suitor because she is not yet ready to copulate. In a typical sequence when a female is in estrus, a male approaches her from behind, sniffs her rear end, mounts, and they copulate. Copulation is short, about 10 seconds or so, after which the male dismounts and the female moves away. Before and during her period of receptivity, a dominant male guards his mate by chasing away subordinates that approach her, but a males' interest in a female lasts only until the end of her short estrus, which appears to end in less than a day.

Courtship and mating is fairly similar in jackrabbits. During the breeding season, male jackrabbits search their home ranges for females. When a male approaches a female, she either boxes him away or rough courtship ensues. Either may begin jumping in the air while the other runs underneath. The jumper often squirts urine on the "jumpee," a behavior called enurination. This may sound disgusting, but in most mammals urine contains useful information: In European rabbits and likely all lagomorphs, chemicals in the urine or associated glands reveal the urinator's sex, dominance status, and reproductive state (see "Do rabbits talk?" in chapter 4).

Next comes a chase, with the pursuing male trying to mount the female, which aggressively rebuffs the male until their interactions stimulate her to receptivity. Finally, after 5 to 20 minutes of courtship, she permits copulation, which lasts only a few seconds. Thereafter she will no longer tolerate the male's advances, although she may copulate again later. One or more males may be involved in chases of a single female, leading to aggressive charges among them that probably establish dominance.

It is also similar in arctic hares, with one interesting twist. Males begin to perform an unusual display at the onset of the breeding season, as described by David Gray: "While the male stands with legs tensed, the long, slender, dark-coloured penis is uncoiled and stretched out along the belly, often extending out beyond the forelegs. This whip-like extension of the penis lasts about 1.5 seconds and is repeated up to 12 times in one bout."

With minor variations, chases by one or more males, fights in which the female is the aggressor, circling, and jumping with enurination followed by very short copulations are characteristic of the courtship behavior of all hares and rabbits in which it has been observed. Courtship and copulation in pikas appears to be a much more subdued affair: a male approaches a female, sniffs her genital area, and mounts. In all lagomorphs, estrus is always brief and, except for the first breeding cycle of the season, occurs just after—and sometimes just before in some hares—the birth of a litter.

All female lagomorphs are believed to be induced ovulators, although this is not definitely known for all species. Induced, or reflex, ovulation means that the stimulation of courtship and copulation is required for a female to ovulate, unlike, for example, human females, who spontaneously ovulate about every 28 days even if no male is present. In European rabbits, ovulation occurs about 8 to 12 hours after copulation. Induced ovulation, which is probably the primitive condition of mammals, is likely an adaptation to ensure no opportunity to breed is missed, which contributes to lagomorphs' generally high rates of reproduction.

The flamboyant courtship behavior of rabbits and hares appears to be energetically expensive. It also seems risky. Rabbits engaged in courtship appear to be less vigilant and their behavior is far from secretive, exposing them to potential predators. Why, then, do they do it? It is possible that the strenuous courtship of rabbits and hares is related to female mate choice. A variety of female mammals prefer to mate with males with characteristics that may increase a female or her offsprings' fitness. These characteristics include a male's age or maturity, dominance status, fertility, genetic compatibility, and even parasite load.

Male rabbits and hares compete for access to females by asserting dominance, either through an established hierarchy, as in European rabbits, or via chases in which a dominant male runs off males that try to court a female he is guarding. At first glance, this appears to give females little choice in the matter, but something more subtle may be going on. Female European rabbits do prefer dominant males; given a choice, females sniff and chin dominant males more than they do subordinates and in the wild the dominant females seldom copulate with subordinates. Dominant males have both higher testosterone levels and larger testes than subordinates, and larger testes means these males produce more sperm. Thus, female rabbits may be choosing males on the basis of fertility. This is important because even an infertile copulation will induce ovulation, which leads to what is called a "pseudopregnancy" that lasts about 2 weeks. Thus an infertile mating may reduce a female's reproductive output.

Courtship interactions, in which females invite males to chase them, scent mark, squirt urine, and display acrobatic leaps, may be the female's way of ensuring that the male she copulates with is truly the best of her choices at the moment, as reversals in male dominance occur regularly. Apart from mere dominance, the male's agility, persistence, and even his ability to expend energy, which may be a sign of health, may all be displayed in his courtship, potentially giving females clues to a potential mate's fitness. Finally, in some species, a female may incite competition when her conspicuous response to a suitor alerts other males to intervene. Where courtship is sometimes violent, as it is when female rabbits and hares fend off courting males, it may be an advantage for her to let another male chase him off instead.

Most lagomorphs are seasonal breeders; even those species or populations with nearly year-round breeding tend to display reproductive peaks—when all or almost all females and males are reproductively active—and valleys when many fewer are. During the nonbreeding season, males' testes are inside their bodies and there is no sperm production because testosterone levels are at baseline. Changes in photoperiod—lengthening days in the late winter or spring—trigger testosterone production and bring males into reproductive condition, usually slightly before changes in day length stimulate estrogen production in females to bring them into reproductive condition. (The reverse happens to end the breeding season, when shorter days return hormones to baseline.)

Other environmental triggers, such as temperature and rainfall that vary from year to year, also modulate the effects of changing daylight, especially among females, so that low temperatures or rainfall may slightly delay the onset of reproductive readiness, and high temperatures or rainfall may advance it. This intricate system helps ensure that breeding begins when there is sufficient new green vegetation to nourish mothers during gestation and lactation and their young when they are weaned and also explains why females are more sensitive to temperature and rainfall than are males, as has been demonstrated in black-tailed jackrabbits, for example. Generally speaking, males need only be concerned with being sexually active when females are, while females have to be able to feed themselves and provide milk to their young. Nonetheless, none of these factors are perfect predictors of future conditions and perhaps as a result, the first litters of the season in some species, as the European hares, are often smaller than subsequent litters. Thus, females may invest less in producing their first offspring of the year because this time is also the riskiest for their survival.

How long are female rabbits pregnant?

Where it is known, gestation length in most pikas is about 30 days. The pygmy rabbit's is a very short 24 days. The gestation length of many cottontails, again where it is know, is similarly in the neighborhood of 30 days, averaging 28 to 29 days in eastern cottontails, 28 to 30 days in the mountain cottontail, 30 days in the Mexican cottontail, and 28 days in the New England cottontail and the brush rabbit. In contrast, the swamp rabbit's gestation length is most often 36 to 38 days, that of the marsh rabbit is reported as 37 to 38 days, the tapeti's is 45 days, and the Venezuelan Lowland rabbit's 35 days. Joseph A. Chapman reported a correlation between latitude and gestation length in New World rabbits, with more southerly species having longer gestation lengths than northern species. However, this relationship may be due to nest site location. Among *Sylvilagus* species, those with belowground nests have shorter gestation periods than those with aboveground nests. This makes sense if young are born at a more advanced state after a longer gestation; it may enable them to better escape predation or to thermoregulate more efficiently in more exposed nests, but more data are needed to test these ideas. Swamp rabbit newborns are more fully furred than, for instance, eastern cottontails, but in both species eyes open by 8 days and young leave the nest at 12 to 15 days (swamp rabbit) or 16 days (eastern cottontail). Among others, bristly, volcano, riverine, and Bunyoro rabbits have relatively long gestations of 35 to 45 days, 39 to 41 days, 35 to 36 days, and 35 days, respectively. Both European rabbits and red rock rabbits have about 30-day gestation periods.

All of the hares, which have precocial young, have longer gestation periods than rabbits, usually around 42 or so days. The snowshoe hare, smallest of the *Lepus* species, has the shortest reported gestation at 34 to 40 days, usually 37 to 38 days, while in the large arctic hare gestation length is about 53 days. The jackrabbits' gestation lengths cluster around 40 to 42 days, while the remaining species' gestation periods range between 37 and 50 days, with European and Cape hares falling toward the shorter part of this range and the mountain and Tolai hares toward the longer.

Two studies have examined the relationship between body weight and gestation length in lagomorphs, with conflicting results. In 1984, Robert K. Swihart, using a data set from 20 species that included pikas as well as values from several populations of some species, found a significant positive correlation between adult female body mass and gestation length. In contrast, in a 2006 study, Emilio Virgós and his colleagues omitted pikas and included a single average body weight and gestation length for each of the 17 species they examined and found no correlation between gestation length and female size. Resolution of this question will await a larger data set that includes more species. The altricial-precocial dichotomy between pikas and rabbits, on the one hand, and hares, on the other, may also confound the analysis.

Where do rabbits give birth?

Pikas give birth either in their burrows or under the talus, depending on the species' typical habitat. Whether females construct special nest chambers in either or both habitats is not known but likely.

A pygmy rabbit about to give birth digs a burrow separate from active residential burrows, usually at the base of dense sagebrush. Natal burrows are inconspicuous, unlike residential burrows. A slanted tunnel about 12 centimeters (5 inches) long and 10 centimeters (4 inches) wide narrows slightly to a 10-centimeter-long entrance to a rounded nesting chamber, which is about the size of a cantaloupe. The chamber is lined with grass and shredded sagebrush bark, then with fur that the mother plucks from her own body. Many rabbits line nests with their own fur; fur on the ventral surface loosens near parturition to facilitate this. The entrance to the tunnel is carefully backfilled with soil and camouflaged with vegetation.

European rabbits similarly excavate a nesting chamber, either within the residential warren or outside of it, but sometimes use a simpler surface nest. Cottontails construct nests, either above or below ground. Eastern cottontail nests are fairly complex. Females dig slanting holes about 10 to 18 centimeters (4 to 7 inches) long, 12 centimeters (5 inches) wide, and 10 to 15 centimeters (4 to 6 inches) deep. The holes are lined with grass, leaves, or soft stems, and then the mother's fur. Nests are found in all kinds of cover, but may be in the open, including on suburban lawns. The desert cottontail's nest is described by mammalogist Joseph A. Chapman as a "pear-shaped excavation," 15 to 25 centimeters (6 to 10 inches) deep, and like that of other cottontails is lined with vegetation and fur. The mountain cottontail's nest is also underground. In contrast, swamp rabbits build an

Rabbit nests may be simple depressions in the underbrush, called forms, or in more elaborate burrows. Photo by Robert K. Lawton, Wikimedia Commons / CC-BY-SA 2.5

aboveground nest of dead weeds wrapped around to create a chamber a few centimeters deep and 15 to 18 centimeters (6 to 7 inches) high and wide, usually at the base of a tree, brush pile, or other object; the marsh rabbit's nest is similar, but usually situated among sedges at the water's edge. These cottontails probably build above ground nests because of the high risk of flooding in their wetland habitats. Riverine, red rock, and Bunyoro rabbits use nests similar to those of eastern cottontails.

Hares, however, because their young can better fend for themselves, create bare-bones nest sites, usually no more than shallow, unlined depressions in the ground very like the forms in which adults rest and located in similar sheltered sites among vegetation. There are a few exceptions. The arctic hare's nest is lined with grass, moss, and fur and set in a well-sheltered spot under or beneath rocks. In the similarly harsh environment of the Alaskan hare, nest sites are sometimes sheltered by thick brush and are sometimes lined; they may also be merely shallow depressions in moss or sedges that are deep enough to leave the young below ground but still exposed to the elements. Jackrabbits also sometimes but not always line their shallow nests with fur or vegetation.

It is commonly argued that the production of precocial young that are fully furred and able to thermoregulate at birth eliminates the need for hares to create more elaborate insulated nests. Moreover, having precocial young is usually viewed as an adaptive strategy that contributes to hares' ecological success. However, recent research by Klaus Hackländer and his colleagues shows that European hare leverets (young hares) exposed to cold or to cool and wet conditions in the first week or two of life expend more energy maintaining their body temperature than they take in by drinking milk. This leaves insufficient energy for growth, movement, or immune responses and may lead to high mortality of leverets in inclement weather.

Rabbits: The Animal Answer Guide

At the same time, other research shows that mothers have a hard time producing enough milk to meet the high energy demands of young. The study authors suggest that production of precocial young may be a constraint, rather than an advantage, caused by hares' inability to dig burrows insulated against extreme weather.

Do rabbits nest at the same time and in the same place every year?

Roughly speaking, yes. Let's start with the second, simpler part of the question. All lagomorphs are relatively sedentary: once they establish a home range or territory they tend to stay there from year to year, provided, of course, that they live from one breeding season to the next. This means that a female will give birth each year in the same home range or territory, although not necessarily in the same exact spot.

As for the same time, the answer is no at a species level, but yes at the population level. To take perhaps the most extreme example, consider the European rabbit. Over its current vast distribution, the timing and the length of the breeding season varies enormously, according to a 2009 review by Zulima Tablado, Eloy Revilla, and Francisco Palomares. For instance, the breeding period ranged from 3 months, March to May, in Washington state; 6 months, March to August, in southern Sweden; 7 months, November to May, in parts of Spain; to year-round in New Zealand. Variation is also seen from year to year in some locations. The most important predictors of the duration and timing of the breeding season are climate and photoperiod.

Climate is influential in several ways. Food tends to be scarce when temperatures are low (or, in hot deserts, very high), and females depend on abundant green, protein-rich vegetation to provide the energy needed for reproduction. Extreme temperatures also increase thermal stress, resulting in poor body condition not conducive to reproduction. Rabbits usually breed in months when day length is long and increasing, as it is during the early spring in the temperate zone, but the day lengths required to breed are affected by food availability and temperature.

The European rabbit is exceptional in the extent and diversity of its range, but all wide-ranging lagomorphs, including eastern cottontails, black-tailed jackrabbits, and snowshoe and mountain hares appear to show significant variation in the timing and duration of the breeding season as well. In the eastern cottontail, the onset of the breeding season is later at higher elevations and higher latitudes. For instance, breeding begins in the first week of January in Alabama and not until the last week of March in Wisconsin. However, there is also year to year variation. The duration of the breeding season also varies from 6 to 7 months in the northern parts of the cottontail's range to all year in the south. As in the European rabbit, this variability is due to the interaction of day length with temperature and food availability.

American pikas begin breeding about a month before snowmelt, which may be as early as March at low elevations or as late as May at high elevations. By mating about a month before the average date of snowmelt, a female produces young during the first flush of vegetation, which provides the nourishment she needs to support the energy demands of lactation. The onset of the breeding season of collared pikas is also timed so that birth coincides with snowmelt. When snowmelt varies from year to year, collared pikas, unlike American pikas, may be able to adjust the onset of the breeding season to match.

Within populations, breeding tends to be highly synchronous, with most adult females mating at about the same time; synchrony is then maintained throughout the breeding season due to postpartum estrus. Unusual among leporids, arctic hares have a very short breeding season and a single litter per year. The onset of breeding is later the farther north the population but is synchronous within populations. For instance, in Newfoundland, fertilization occurs within about a week in mid-April, and all litters are born in late May or early June. Where the growing season is very short, as it is in northern climates, timing is tricky because young need enough time to grow big enough to endure the harsh winter conditions.

In less seasonal environments at low latitudes, where daylight is more or less constant through the year, breeding of Cape hares living near the equator and red rock, forest, and volcano rabbits, and others is usually year-round, although it may be influenced by seasonal changes in rainfall.

How many babies do rabbits have?

In general among mammals litter size is correlated with body size and longevity: species that are small and short-lived have larger litters than larger, longer-lived ones. Simply put, if you don't have long to live, you need to pump out as babies as possible in the time available to you. If you're likely to live a long time, you can take your time producing babies. In broad comparisons between rabbits and hares, this holds true. Rabbits are smaller and live shorter lives and tend to have larger litters than hares. But body size in pikas appears to be less important than their longevity. Pikas that live in talus generally live longer than those living in burrows. Yet, the talus-living pikas have small litters and tend to have only one per season. The burrowing pikas have several large litters each breeding season. So the talus-living American pika starts out with litters of from one to five but because of deaths before weaning, including in utero, most females end up weaning about two young per year. But these young may live to be

6 years old. In collared pikas, females may wean only one juvenile per season, from litters averaging two or three young. In contrast, plateau pikas produce at least three litters of four to eight; most are weaned successfully, but few live to the next year.

Clearly other factors also affect litter size and litters per season, including where newborn are on the altricial-precocial spectrum. For instance, a hare's smaller litters may be due in part to their babies being larger at birth than those of rabbits. A recent comparative study involving 11 orders of mammals, including lagomorphs, showed that both litter size and total litter mass are related to how newborn size affects survival to weaning. Mammals whose babies are born and reared in nests or burrows, where size doesn't usually greatly affect survival from birth to weaning, have large litters of altricial babies, as rabbits do, while those whose babies are raised in the open, exposed to predation and the elements where larger body size is an advantage for escaping predators and thermoregulation, like hares, have smaller litters of precocial young.

The length of the breeding season is also important—a short breeding season correlates with larger and fewer litters. Across the hares, for instance, species (and populations of species) living in the far north have one litter of six to eight young while those at the equator may have eight litters of one or two, "giving a surprisingly constant annual production of about 10 young per female per year," according to John Flux and Renate Angermann.

Similarly, among cottontails, litter size varies with latitude and altitude, within and among species. Average litter size in eastern cottontails varies from 5.6 in Illinois to 3.1 in Georgia, and 2.0 in South American cottontails, with a species-wide range of 1 to 12 per litter in up to eight litters per year. Unlike the hares, however, the eastern cottontail's annual productivity is highly variable, ranging from about 6 to as many as 35 young per female per year, and the eastern cottontail is the most prolific species in this genus. The New England cottontail, at the northern edge of cottontail range, has two or three litters averaging 5.2 young per year. The South American tapeti's average litter size is 1.5, with about nine litters per year, while that of the Venezuelan lowland rabbit is 2.6, for a total of about 17 per year. Volcano rabbit's litter size is small, 2.1, but this species may breed all or most of the year. Pygmy rabbits have litters of about 6 young, with a maximum of three litters a year.

These generalizations hide a significant amount of variation within lagomorph species. For instance, in some seasonal breeders with multiple litters, such as most cottontails, the snowshoe hare, the plateau pika (but not the American pika), and others, the first and last litters of the season tend to be smaller than those in the middle, because food resources and optimal mild temperatures peak in mid-season. The first litters of alpine pikas are smaller than the second ones. In many other species, including eastern cottontails, European rabbits, and some pikas, younger, juvenile females tend to have smaller litters than older ones; there is some evidence that litter size tapers off among the oldest females. Over its current range, the European rabbit exhibits average litter sizes that vary from 3.2 in parts of Spain to 7.3 in Washington state, with most populations averaging a litter size of about 5. In the Tablado study mentioned above, litter size was not correlated with the length of the breeding season but was related to average adult body weight; larger mountain hares, measured by hind foot length, also produce larger litters than smaller hares do. In another study comparing rabbit populations in different climate regions of Australia, average annual litters ranged from 4.5 to 5.7 litters of from 2.9 to 5 young, and total young per year per female from about 14 to 28. In snowshoe hares, average litter sizes vary from year to year, depending on where the population is in its 10-year cycle: litter sizes are smaller during the decline and low years and larger during the increase and peak years.

A detailed study of European hares found seasonal differences in litter size, with mothers having 1 or 2 young between November and March, and 2 or 3 between April and October. Babies born in litters of 1 or 2 were larger, about 123 and 112 grams (4.3 and 3.9 ounces), respectively, that those born in litters of 3 and 4, about 105 and 95 grams (3.7 and 3.3 ounces), respectively. This suggests that during the colder parts of the year, it is better to produce fewer, larger babies better able to meet the demands of thermoregulation, and more, smaller babies under more benign conditions.

Are all babies in a rabbit's nest full siblings?

The mating system of many lagomorphs is described as promiscuous, meaning that both males and females may copulate with more than one mate during a single period of estrus. Wherever this is the case, there is some likelihood of the young in a single litter being fathered by more than one male. In many leporids, a male dominance hierarchy exists around estrous females; the dominant male achieves most of the copulations but subordinates are not totally excluded. This has been conclusively demonstrated in brush rabbits, in which genetic analysis showed that some females had litters of mixed paternity.

Among snowshoe hares, genetic analysis showed that litters of mixed paternity occurred, but less often than expected if mating was completely promiscuous, suggesting multiple successful copulations by either males or females were not common. In a study of European hares, about 10% of litters were sired by more than one male.

In some species, social organization may largely preclude females from mating with more than one male. In the fiercely territorial American pika, for instance, opportunities for a female to encounter and copulate with a male other than her nearest neighbor may be limited; males also guard females. However, especially where a male tries to guard two females, a female may have a chance to copulate with other males on occasion. Female alpine and northern pikas, in which pairs defend a territory, are also reported to regularly copulate with several mates.

In the more social plateau pika, family groups may be monogamous, polygamous (one male and more than one female), polyandrous (one female and more than one male), or polygyandrous (multiple males and multiple females). All of the males in polyandrous and polygyandrous groups may sire some of the young in this species' large litters. Females usually copulate with all of the adult males in their family during estrus, although the dominant male in multimale families sires more offspring than subordinates do. Rarely, a female even copulates with a male outside of her family.

The question of why a female should mate with multiple males intrigues many scientists. It is well established that, in most species, male reproductive success is increased by mating and producing young with as many females as possible. In general, for males the low costs of seeking and inseminating multiple partners are more than offset by the potential benefits of producing more young. The benefits to females of multiple matings are less obvious, while the potential costs include increased exposure to predation and increased risk of disease transmission.

One hypothesis to account for female promiscuity is called the genetic incompatibility avoidance hypothesis. The idea is that by copulating with multiple males, a female reduces the costs of having her eggs fertilized by sperm that are a poor genetic match. Genetic incompatibility often results in the death of developing embryos, such that the female has wasted her energy investment in producing eggs and supporting embryos. This hypothesis assumes that polyandrous females can somehow filter out incompatible sperm before fertilization or quickly reabsorb any embryos sired by the incompatible sperm, leaving only compatible sperm to sire embryos that develop to full term. Where litters of mixed paternity are born, this hypothesis suggests that either these postcopulatory filters failed or that the sperm of all of the males a female copulated with was compatible. A recent comparative study by Paula Stockley provided support for the genetic incompatibility avoidance hypothesis, showing that across a diverse range of mammals, including a comparison of European rabbits and eastern cottontails, females in more frequently polyandrous species showed lower rates of early reproductive failure than females less often polyandrous. So why aren't all female mammals promiscuous? As in the American pika, monogamous females may simply have limited opportunities to mate with more than one male.

Some European hares also exhibit a phenomenon called *"superfetation,"* in which females begin a new pregnancy near the end of an existing one. They can do this because they have two uteri. In a study of captive European hares, for instance, successful copulations were observed after about 34 days into the typical gestation period of about 41 days. In this study, 85% of pregnant hares mated and ovulated at the end of pregnancy and among these, 59% exhibited superfetation. This occurs in wild hares, but apparently rarely, and is more common in captive hares when males and females are housed together continuously. The occurrence of superfetation is usually inferred from observations of a shortened interval between consecutive parturitions. However, one study of captive hares suggested that hormonal changes associated with copulation and induced ovulation during late pregnancy may precipitate premature births so interbirth intervals may not be reliable indicators of superfetation. Based on the method of dating interbirth intervals though, superfetation is also reported in Japanese hares.

Do rabbits care for their young?

Not so much. Except for some pika species, fathers contribute very little or nothing to caring for young and even mothers practice what is called absentee-maternal care. In other words, mothers best care for their young by staying away from them as much as possible. This is likely a strategy for reducing predation. Mothers have little ability to defend young from predators, although they may approach a baby that screams in fright. So mothers avoid as much as possible leading predators to their babies' hiding place by leaving their babies soon after parturition and returning only infrequently to nurse for a very short time. And even then, mothers appear to do little or nothing to facilitate their babies' suckling other than approaching them and then sitting or hunching over to present their nipples.

Robyn Hudson and her colleagues and students have studied mother-young relations in domestic European rabbits for many years; many of their findings are summarized in a paper in the book *Lagomorph Biology*, on which the following is based. Rabbits usually give birth during the day, their typical rest period. Parturition, even of very large litters, takes only 10 or 15 minutes. A mother sometimes licks the neonates as they are born and always quickly eats the placentas. As soon as the last baby is born, the mother leaves, covering up the entrance to the nest chambers as she departs. Thereafter, she visits the nest just once every 24 hours, at night, for 3 or 4 minutes.

When the mother arrives at the nest she immediately hunches over the young in a nursing posture and the babies rear up, find a nipple, and suckle rapidly. Although the suckling bout may not last long, significant amounts of milk are obtained only during the second minute of the bout. A baby

Typical of lagomorphs, the mother of these brush rabbit young will visit just once a day to nurse them.
Photo © Chris Wemmer

may gain up to a third of its body weight during this short session. Other than perhaps a little licking, the mother doesn't interact with her babies at all and when the babies' time is up, she jumps away, defecates a few hard feces, and leaves. And that's it for another 24 hours even if the competition among young meant that some got no milk at all.

Like athletes warming up before a game, the rabbit babies start getting primed for their nightly feed about an hour before their mother gets there. They become more active, respond more readily to touch, and squirm out from beneath nesting material. Their body temperature also rises. This means that they ready to fight for their meal the instant mother moves into position. What controls the circadian rhythm of this anticipatory warm-up session is unclear, but it's not just due to hunger or any empty stomach. Litters that miss a meal—babies can survive 48 hours without feeding— similarly warm up 24 hours later.

Babies use odor cues to find and latch onto a nipple. The "nipple-search pheromone" is present in milk (it is also emitted during late pregnancy and, in lesser amounts, during estrus) and appears essential to eliciting the pup's search-and-suckle behavior.

A mother's milk production begins to taper off when pups are about 20 days old. If a mother is pregnant, she then abruptly stops visiting her young when they are about 26 days old—one night she's there, the next she's gone forever. A female who is not pregnant may nurse a few days longer, weaning her young at 30 days or so, but there's the same sudden abandonment of babies to their fate. In the wild, rabbit mothers close the entrance to the nest burrow when they leave each night. Once young begin to leave the nest in the evenings when they are a few weeks old, they later assemble at the nest site to await their mother's visit. This continues until they are weaned.

With essentially no mothering, rabbit babies both compete with and partly depend upon their siblings. As mentioned above, babies compete for nipples. The heaviest babies get more milk than lighter ones, leading to higher probability of surviving—as many as 20% of babies starve to death in their first week of life. They also have the advantage of being heavier when they are weaned than their smaller littermates. But having no littermates is not ideal, even though littermates are competitors. A single pup doesn't stimulate its mother enough for her to stand still for nursing or to release the hormone oxytocin that is needed for milk letdown.

Heavier babies also do better in the competition for the warmest spot in the nest—which is a place where a baby has as much of its body surface touching other babies as possible. In the best situation, a baby would be lying on another one, while others covered its back and rested along its sides. Most of babies' active time is devoted to climbing over, under, and around each other to achieve the warmest position. Again, heavier babies do better than lighter ones and babies with a high "huddle index" are better at converting milk to body weight. And, as with nursing, despite the competition, having littermates is important to keeping warm. Even the babies in litters of only two have higher body temperatures and grow better than singletons.

With a few exceptions, the pattern of maternal care is broadly similar in other lagomorphs. Except possibly for pygmy rabbits, lagomorphs mothers don't retrieve their young or provide any help to them beyond just showing up. Like European rabbits, cottontail mothers simply visit their young once in the evening for short nursing bouts; they open up the entrance to the nest and hunch over it—young come to them—then cover the entrance again when they abruptly leave. Nursing bouts generally last less than 10 minutes. Among hares, mothers also visit to nurse their young only once a day, usually just after dark, although arctic hare mothers stay with young for the first 2 to 3 days and try to defend them should danger lurk.

Precocial hares differ from rabbits in leaving the nest very soon after birth and in snowshoe, European, and arctic hares and others go to their own separate hiding places. In spite of having their own hiding places, young hares still reassemble nightly near their birth site to be ready for their mother's daily visit. In arctic hares, young begin to arrive near the nursing site 20 to 90 minutes before their mother arrives, 15 minutes earlier in European hares, and 5 to 10 minutes in snowshoe hares. This sug-

Huddling with littermates is important to young rabbits' ability to stay warm. Photo by Ben23, Wikimedia Commons / CC-BY-SA 3.0

gests that hare young also follow a circadian activity pattern that anticipates the mothers' visit. Some research on captive Cape hares suggests that hare may find their birth site area using olfactory cues, perhaps from recognizing their own body scent rather than that of their mother.

In European hares, young are weaned at about 30 days, longer if the mother isn't pregnant. But young begin to eat vegetation at about 8 days and milk is probably only a supplement to vegetation by about 17 days. The length of the daily nursing bout in the young's first week varies from 2 to as long as 8 minutes but most often less than 6 minutes; by the third and fourth week 2 to 3 minutes is the norm. When some mothers nurse their young longer, say, 5 to 6 weeks, nursing bouts are even shorter. As in rabbits, mothers wean young by simply not showing up to nurse. Hares show one additional maternal behavior not observed in rabbits. At the end of the nursing bout, the mother licks the leverets' genital region and probably drinks the leverets' urine. This too is most likely antipredator device, so the scent of urine doesn't reveal the gathering site to predators.

The solitary and territorial talus-living pikas that have been studied differ little from the leporids in their parental care, although nursing may be more frequent. For example, in the alpine pika, fathers play no role and mothers visit the nest to nurse young every 2 hours for about 10 minutes. Mothers don't retrieve young and no other maternal behaviors are observed. When young are about 6 weeks old, the mother no longer visits them to nurse, and the youngster's attempts to nurse are often met with maternal aggression. The maternal behavior of collared pikas is similar, although weaning occurs as early as 3 or 4 weeks of age.

In the more social burrowing pikas, such as the plateau pika, females exhibit the typical absentee-mother pattern of care. Observations in a recent study show that both mothers and fathers with young spend on average almost 90% of their time on the surface during the 12 daylight hours, suggesting short, infrequent nursing bouts in these pikas too. Surprisingly, survival rate of young from birth to 15 days was greater the more time their mothers spent above ground, perhaps because more time above ground means greater food intake and thus better ability to meet the nutritional needs of her young. Once young are active above ground, at 15 to 45 days, their survival rates were positively correlated with the amount of male above ground activity. Males (and much less frequently, females) during this period are attentive to the juveniles, showing friendly behaviors such as sitting together, grooming, and mouth and nose rubbing. Males also are more vigilant than females, and this in particular may enhance juvenile survival by alerting them to approaching predators in time to retreat underground.

How fast do rabbits grow?

On average, lagomorphs grow relatively more rapidly than many mammals, and pikas grow relatively more rapidly than other lagomorphs. From a birth weight averaging about 12 grams (0.4 ounces), an American pika reaches minimum adult weight of about 120 grams (4.2 ounces) within 3 months. Such rapid growth is necessary because young must reach adult size, establish a territory, and collect hay piles before winter sets in (see "Do rabbits ever store their food" in chapter 7). With a shorter season, the more northerly collared pika grows even faster than the American pika. These talus-living pikas do not breed until they are yearlings, but in burrowing steppe pikas, females are sexually mature at 4 to 5 weeks of age! American pika young don't open their eyes until 9 days after birth but can survive without their mother by 18 days, although they are typically weaned at 21 to 28 days. By 28 days, young are pretty much on their own, aggressively avoiding contact with their mother and siblings.

Eastern cottontails, born at about 40 grams (1.5 ounces), triple their weight by about 2 weeks of age and reach 800 grams (28 ounces) by 6 months, a twenty-fold increase. Growth of different parts of the body is not uniform. For instance, hind-foot length reaches adult size at 3 months, probably because this is key to running efficiency. All skeletal growth is complete by 5 months but full adult weight is not reached until about 10 months. Nonetheless, females may breed as about 3-month-old juveniles. All cottontails achieve various developmental milestones at roughly the same ages: eyes open at 4 to 10 days, leave the nest at 12 to 16 days, are weaned by about day 30, and disperse soon afterward.

Hares generally reach minimum adult weight by 3 or 4 months of age.

Born with little fur and closed eyes, cottontails grow quickly to resemble miniature adults. Photo by Gary M. Stolz, USFWS

Alaskan hares, for instance, go from 100 grams (3.5 ounces) at birth to 3.9 kilograms (8.5 pounds) in just over 100 days. As in the cottontails, growth rates vary among different parts of the body. In black-tailed jackrabbits, for instance, the skeleton is about half of adult size at 1 month, but its weight is only 13% of adult weight. At 2.5 months, ear length, hind-foot length, and skeleton reach more than 90% of adult size, but weight is only 65%. Hares usually do not breed until the late winter or spring after their birth.

As with other life history traits of lagomorphs, however, there is significant variability in growth rates. Mountain hare young reach adult size in about 4 months, but rates of growth vary with the subspecies and over the breeding season. Young born early in the breeding season have more time to grow and so end up larger as they head into winter, while young born later grow faster but for less time. Growth rates of European hares are influenced by litter size, with young from smaller litters growing faster than those from larger ones. Siblings have to share the finite supply of their mother's milk, so in large litters the share of milk each receives is smaller.

In the well-studied domestic and European rabbits, growth rates are significantly affected by litter size and also by birth weight, such that heavier babies grow faster than lighter ones, perhaps because they compete better for their mother's milk. Growth rates in European rabbits are also influenced by temperature, with babies growing faster at higher temperatures than lower ones. At lower temperatures, babies must devote more energy to thermoregulation, leaving less to allocate to growth. This relationship is further complicated by litter size because babies in larger litters, thanks to their huddling behavior, expend less energy on thermoregulation. Thus, a recent study by Heiko G. Rödel and his colleagues showed that babies in litters of three grew relatively faster at lower temperatures than pups in litters of two, even though more siblings were competing for milk.

The rapid growth of lagomorphs before weaning is fueled by their mothers' energy-rich milk that is high in both lipids (fat) and protein and low in lactose (sugar). For instance, eastern cottontail milk is about 14% fat and 15% protein, domestic rabbit milk is 12 to 13% percent fat and 10 to 12% protein, and that of European hares has been shown to vary with diet: mothers on a high-fat diet produced milk with 26% fat and 12% protein versus mothers on a low-fat diet whose milk was 20% fat and 15% protein. To put this in perspective, domestic rabbit milk is twice as dense with fats and three times as dense in protein than is cow's milk.

How can you tell the age of a rabbit?

Once a lagomorph reaches adult size and coloration at about 1 year, it is very difficult to tell its age by external appearance or any other feature. The ages of juveniles can be estimated by developing growth curves; this involves frequent weighing (or measuring) of known-age young from as near to their birth as possible to the point when their weight levels off as they reach adulthood. With such a growth curve, one can estimate the age of an individual of unknown birth date by matching its weight to the right point on the curve. Obviously, it is difficult to develop growth curves; moreover, growth curves can only be relatively accurately applied when the animals come from the same population—not just the same species—because growth rates as well as average adult size vary between populations.

Scientists have developed several ways to distinguish between individuals 1 year old or less and individuals older than 1 year. For instance, the weight of the eye lens predicts the age of cottontails and hares younger than 1 year old and in arctic hares can accurately predict the age of older animals as well. Among young cottontails, the growth of some skull characteristics is useful for placing young under 170 days old in broad age categories. One technique that has been applied with some success to determine the ages of adult hares and rabbits is to measure growth lines in the lower jaw bones. Like tree rings, the number of lines indicates the age of the individual. Of course, the animals must be dead to use these measures.

All this means that the best the average person spotting a rabbit in the field could do is to guess, based on comparative size, that the animal was babyish, youngish, or adult.

How long do rabbits live?

There is a general relationship among mammals between body size and longevity: larger species live longer than smaller ones. For their body size, lagomorphs are short-lived, although their potential life spans are much

Young pygmy rabbits emerge from their natal burrows at about 2 weeks of age, ready to eat the sagebrush that is the mainstay of the species' diet. Photo © Jim Witham

longer than their typical life spans. For example, the potential life span of an eastern cottontail is estimated to be 10 years, while the average life span is about 15 months. Similarly, the longest life span known for a wild European rabbit is 7.6 years, and well-tended house rabbits can live 9 to 10 years, but 80% or more of wild rabbits die before 3 months of age, and about half of the adult rabbits alive at the beginning of one year are dead by the end of that year. This means that vanishingly few wild rabbits live to the old age they are potentially capable of reaching.

Among all of the leporids, adult mortality rates are generally lower than juvenile mortality rates. A study of arctic hares, for instance, showed that the annual adult survival rate was 78%, but only 15% of young survived their first year. Similarly, a 5-year study of mountain hares in Sweden reported that 85% of young died between 14 days of age and the beginning of the breeding season the following spring, while annual adult survival ranged from 42 to 88%. Studies in other parts of this species' range report adult survival rates as low as 6 to 44%. In various studies in different locations, black-tailed jackrabbit juvenile mortality averaged 59% (range 24 to 71%) and 63% (range 35 to 67%), with one estimate of 91%. Over 8 years, adult mortality ranged from 9 to 87%, but averaged about 55%. The maximum age reached by black-tailed jackrabbits is estimated to be 7 years.

In the long-term study of snowshoe hares in Kluane in Yukon Territory of Canada, survival rates of young until 30 days ranged from zero to 73%, depending on whether the population was in the declining or increasing phase of its 10-year cycle. Likewise, annual adult survival ranged from 0.5 to 32%. The oldest hares found in this study were 6 years old, but these

Young lagomorphs, like this juvenile desert cottontail, suffer from high rates of mortality. Photo by H. Cheng, Wikimedia Commons / CC-BY-SA 3.0

hares were living in enclosures that excluded predators; where predators were not excluded, the oldest hares were 5 years old.

Survival rates of young rabbits and hares is typically lower than that of adults because young are more susceptible to predation, partly because their running abilities are not yet fully developed. Younger, smaller animals are also susceptible to a wider range of predators; for instance squirrels can only eat baby snowshoe hares that are less than 14 days old, snakes and corvids (crows and their relatives) prey only on very young cottontails, and raccoons and skunks take newborn but not adult jackrabbits. Inexperienced youngsters may also be less adept at finding escape routes and hiding places in unfamiliar surroundings, while experienced adults know their home ranges more intimately. Young also die of exposure, starvation, drowning, and disease at higher rates than adults do.

In some species, such as the mountain hare, adult males tend to have higher mortality rates than females. Males generally expand their home ranges during the breeding season in order to find receptive females, so the risk of predation may be greater when they venture into unfamiliar areas.

Pikas show an interesting contrast in survival rates between the talus-living species and the burrowing species, exemplified by America and plateau pikas, respectively. American pikas may live as long as 7 years, and annual mortality rates are 37 to 56%. Plateau pikas have very high rates of mortality. Of 324 young marked in one study, fewer than 16% survived to breed in the year after their birth and only 1.5% lived to breed in their second year. The reason for this dichotomy is not known, but it is correlated with differences in the two types of pikas' reproductive strategies: during the breeding season, plateau pikas have more and larger litters than American pikas.

Rabbits: The Animal Answer Guide

Chapter 7

Rabbit Foods
and Feeding

What do rabbits eat?

Most people asked this question would answer carrots, based on the near-universal association of rabbits and carrots in popular depictions, such as Bugs Bunny who is rarely seen without a carrot in his mouth. And it's true that rabbits and hares like carrots, as well as lettuce and other veggies that Peter Rabbit filched from Mr. McGregor's garden. In agricultural areas of Europe, where European hares are abundant (but declining), in some seasons they prefer to dine on cultivated crops, such as wheat, barley, and alfalfa, and on the sugar beets and carrots that hunters set out as bait. It appears that all rabbits and hares, and perhaps pikas as well, will partake of human food crops when they are available.

Lagomorphs are all vegetarians, with a few unusual exceptions. The arctic hare is known to eat some meat, including fish that they find in traps baited for carnivores and the stomach contents of caribou whose guts have been exposed by a carnivore. They have also been observed visiting garbage dumps in winter. This doesn't seem to be a regular occurrence, however, so it's not clear whether this hare's taste for flesh is important to survival.

Meat-eating does appear to be critical for collared pikas living in an isolated population in the Yukon of Canada, which was discovered David Hik. These pikas eke out a hard living on nunataks, small islands of rock that jut out of the vast glacier of the St. Elias Mountains. Small meadows at the tips of nunataks support some plants, but all are in short supply. To supplement their meager plant fare, these pikas have taken to eating the brains of birds. During their spring and fall migrations, songbirds run into storms that smash them down onto the glacier, where they often die. Others succumb to exhaustion and fall to the ice. Pikas fetch the tiny carcasses and gnaw little holes in the back of the head to extract brains, which are high in protein and fat. This nutrient boost may be just what these pikas need to prepare for the demands of reproduction in the spring and for the long bleak winters in the fall. The pikas even store some carcasses in the hay piles they rely on for winter sustenance (see "Do rabbits ever store their food?" later in this chapter).

Other than that, lagomorphs are what scientists call generalist herbivores. They eat a wide variety of grasses, forbs (most weeds and wildflowers such as dandelions and clover are forbs), leaves and twigs of shrubs, flowers, bark, and other plant material, with the occasional addition of berries, seeds or nuts, lichen, and fungi. Individual species vary in their preferences for and consumption of various plants and plant parts, depending on what is available, and in many cases these vary considerably over the course of a year. As a generalization, lagomorphs graze on fresh grasses and forbs when they are available, such as in the spring and summer in the temperate zone, and switch to browsing on shrubs, twigs, and bark when green food largely disappears in the winter.

Most lagomorphs eat a large number of species of vegetation, but this doesn't mean they eat anything and everything. For instance, in a study of European hares by Thomas Reichlin, Erich Klansek, and Klaus Hackländer, some 164 plant species were available, but only about 50 were found in the hares' stomach contents. Further, depending on the season, just three species composed up to about 90% of the stomach contents. Similarly, Pippa Seccombe-Hett and Roy Turkington found that snowshoe hares ate just 10 of 30 plant species available to them in a large experimental enclosure, and 5 of these 10 formed more than 90% of the stomach contents in the early summer.

Lagomorphs select particular plants or types of plants (such as grasses versus forbs) to balance their need for protein, energy, fiber, water, fat, and other nutrients. They also have to avoid being poisoned. Just as lagomorphs have defenses to protect them from predation, many plants also have defenses again being eaten. Thorns, for example, evolved to make it more difficult for herbivores to munch on leaves. More often, plants use chemical defenses, called "plant secondary compounds." These are chemicals that are more or less toxic to animals that eat them, often impairing digestion or some other physiological process rather than killing an animal outright. Many of the herbs and spices we use to make our meals more appealing, such as chili peppers, cinnamon, and rosemary, get their zing from these secondary compounds but in amounts too small to be toxic to people. Likewise, addictive chemicals such as nicotine, caffeine, and cocaine are plant secondary compounds, as are the tannins in tea. Salicin, a plant secondary compound found in the bark and leaves of willows, is the basis of salicylic acid, the effective ingredient in aspirin.

Grasses do not contain secondary compounds, which, along with their high water content, may be one reason they may be preferred by most lagomorphs. Similarly, the domesticated crops many lagomorphs enjoy have lost their toxicity, making them safer and more palatable to both humans and wildlife.

Wild forbs, shrubs, and trees, however, are full of secondary compounds, creating a challenge for herbivores, which have evolved various strategies to cope with potentially toxic diets. One strategy is to eat small amounts of a variety of species with different secondary compounds. Different classes of secondary compounds are detoxified using different metabolic pathways, so the idea is to mix poisons so as not to overload any one of these pathways; it is also possible that the chemicals found in one plant may counteract the toxic effects of the chemicals in another. Another strategy is to amp up the body's ability to detoxify one or more types of plant secondary compounds. Some detoxification occurs in the liver, but other processes reduce absorption in the digestion track. Another is to regulate meal size and the interval between meals so an animal stops feeding on a particular plant when toxins in the blood each some level and doesn't start again until toxins decline. Finally, they may somehow manipulate food items in such a way as to reduce the amount of toxins before they eat them. The food plants that American pikas store in their hay piles, for instance, are high in toxins, but the toxins degrade during storage.

Various plant secondary compounds have been shown to deter or reduce browsing in snowshoe hares. Camphor, for instance, turns snowshoe hares away from feeding on juvenile white spruce, while adult white spruce, with about one-fourth the camphor content, is highly preferred. Another chemical protects Labrador tea, a small evergreen plant, from snowshoe hare grazing. However, most of the plants that snowshoe hares eat are heavily defended by plant secondary compounds, and they eat these plants even when less well-defended plants are available. These hares appear to be selecting plants for their nutritional content and manage the toxins by mixing their poisons—in summer eating plants with high levels of tannins and plants with high level of alkaloids at the same time or in winter mixing those with resins, such as birch, and those with phenols, such as spruce.

However, snowshoe hares do prefer older plant parts of quaking aspen, balsam poplar, paper birch, and green alder, which have lower concentrations of their various plant secondary compounds. Similarly, mountain hares prefer older parts of willow to younger ones.

Snowshoe hare herbivory also strongly affects the evolved chemical defenses of the woody plants it eats. For instance, John P. Bryant and his colleagues found that juvenile paper birch produce three orders of magnitude the amount of resins where they are heavily browsed by snowshoe hared in Alaska, compared with juvenile paper birch in eastern North America, where browsing by these hares is less intense. Aspen are similarly strongly defended by chemicals in Alaska and less so in eastern North America. Selective browsing by snowshoe hares and other herbivores on more palatable woody plants can also affect the species composition of a forest, so that in some cases the plants most strongly defended by toxins come to dominate.

One consequence of herbivores meeting the challenges posed by plant chemical defenses, therefore, is that few herbivores—only about 1%—depend on particular plants for their diets. The pygmy rabbit is an exception. Sagebrush makes up about 50% of its summer diet and 99% of its winter diet. Yet sagebrush, which is laced with bitter-tasting volatile oils called monoterpenes among other chemicals, causes problems even when eaten in small amounts by mice, sheep, and cattle. A detailed study by Lisa A. Shipley and her colleagues compared the food preferences of pygmy rabbits and eastern cottontails, which may include some sagebrush in their diets when they occupy sagebrush habitats. The results showed that both species, given a choice, preferred pelleted rabbit food over sagebrush, but pygmy rabbits' diet still included 8% sagebrush while the cottontails' was less than 1%. Pygmy rabbits could meet their daily requirements eating only sagebrush, although only barely, while eastern cottontails ate only enough sagebrush to meet about two-thirds of their requirements. This suggests that pygmy rabbits have a greater ability to deal with monoterpene toxins than eastern cottontails do. And one way they might do this is by getting rid of the toxins almost immediately. Research from the 1980s showed that significant amounts of the volatile oils are dispersed—volatilized—during chewing so little reach the rest of the digestive track or the liver. Whether eastern cottontails do this to a lesser extent or not at all is not known.

The effects of eating phenol-laden birch on mountain hares, which eat a lot of birch in winter, and on European hares, which eat very little birch, were compared in another study. Both types of hares ate less and showed a reduced ability to digest the protein in birch as phenols increased, but European hares lost serious amounts of sodium in their urine while mountain hares did not. This suggests that the mountain hares have a greater ability to detoxify phenols than European hares have.

Much remains to be learned about how lagomorphs cope with the plant secondary compounds that defend many if not most of the nongrass plants they eat. There is a great deal of information about a set of enzymes, referred to as P450s, that work to detoxify chemicals through oxidation, primarily in the liver. This topic was reviewed M. Denise Dearing, a leader in this field for many years. All vertebrates appear to have these enzymes, but they vary among species. These enzymes are well studied in people and rats. Two studies show that the P450 enzymes of a marsupial leaf-eater whose diet is full of terpenes are better able to detoxify terpenes than the enzymes of people and rats. It would be interesting to learn how the P450 enzymes of lagomorphs compare. It has been shown that there has been positive selection for gene sequence sites coding for these enzymes in rabbits, most likely to maintain mutations that increase the number of chemical that can be detoxified.

Another detoxification method involves a molecule that binds to a toxin under the influence of another set of enzymes called UGTs (for UDP-glucuronosyltransferase); this binding makes the toxin easily excreted in urine. Different UGTs may act on different toxins or classes of toxins. A recent study showed that rabbits have a larger number of genes coding for UGTs than do people and rats, suggesting that the diverse toxins that lagomorphs find in their food selects for a greater array of enzymes to counteract them.

How much do rabbits chew their food?

Chewing, or mastication, begins the process of digestion by mechanically cutting or grinding food into a form that can be swallowed and through the action of enzymes in saliva that begin the chemical breakdown of the food. How much a rabbit chews depends on the food item—tender grass needs less chewing than a tough twig. In one highly artificial laboratory study, domestic rabbits spent about 2% of the day chewing, and an arctic hare was reported to have spent about 3 minutes eating a 25-centimeter- (10-inch-) long twig. But mastication has been studied in detail only in domestic rabbits in a laboratory setting.

Mastication is surprising complex, involving the jaw muscles that power the teeth as well as muscles controlling the lips, tongue, and cheeks. Mastication begins with the rabbit using its incisors to snip off a piece of vegetation that is then moved by the tongue to rest between the upper and lower molars on one side of the jaw. Powered primarily by large masseter muscles, the lower jaw closes and moves from side to side so the lower molars grind the food against the upper ones. The sharp ridges around the edges and across the center of the cheek teeth help shred the food as it is ground. Tongue, cheek, and lips muscles help to keep the food between the jaws. Lagomorphs chew on only one side of the jaw at a time: the distance between the teeth on either side of the upper jaw is greater than the distance between the teeth in the lower jaw, so occlusion between upper and lower teeth is only possible on one side of the jaw at a time.

The pattern or sequence of mastication in rabbits just described is highly stereotyped and has been divided into three types. Type I, called the preparatory sequence, is when the lower jaw moves up and down so the incisors can cut off manageable bites of food. Type II consists of a fast opening phase, a fast closing phase, and a slow closing phase when the food is actually ground between the teeth. Type III involves jaw movements that bring the food to the back of the mouth for swallowing. This pattern is not affected by the texture of the food being chewed. Rabbits use the same sequence whether they are eating a twig or a soft piece of grass, but the force that the teeth apply during the crushing phases increases as the hardness of the food item does. Rabbits exert about twice as much force to chew dry pellets than they do to chew carrots. The secretion of saliva is also greater when rabbits chew pellets compared with when they eat carrots, which are mostly water. Salivary secretion also varies with the side of the mouth being used, so as food is shifted from one side of the jaw to the other, the flow of the secretion of the glands on either side of the tongue stops and starts.

The relative ability of individuals to chew hard food is not fixed but is the result of exercise. In laboratory studies, young rabbits fed a tough diet developed larger masseter muscles and thus could exert greater bite force that those fed a soft diet. There is also sexual dimorphism in the development of masseter muscles in rabbits: under the influence of testosterone, the masseter muscles of males grow larger than those of females, although why this should be the case is not known. Surprisingly, rabbits appear to need practice to perfect chewing. Chewing movements develop from sucking movements in young rabbits and by about 2 weeks of age their chewing movements are similar but not identical to those of adults. It takes about 2 weeks for their masseter muscles to develop and for chewing to be as coordinated as it is in adults.

There are some small but significant differences between the mastication of European hares and European rabbits, described by lagomorph biologist Philip Stott. Rabbits cut and crush twigs more completely than hares do, using the side to side movement between the molars described above. Hares appear to add a movement that strips the twig so pieces are longer and thinner than the pieces chewed by rabbits. This may expose the carbohydrates within the twig, which are more abundant within some parts of the twig than others, so they can be more quickly absorbed during digestion. This also allows for the indigestible, fibrous parts of the twigs, which also contain more tannins, to pass through the digestive tract more quickly.

Rabbits: The Animal Answer Guide

How do rabbits find food?

The home ranges of lagomorphs usually contain all of the food that they need to survive throughout the year. Pikas and rabbits rarely travel very far from their burrows or other forms of shelter—on the order of a few or a few tens of meters (1 meter equals about 1 yard). Hares may travel farther from shelter to open feeding grounds. Black-tailed jackrabbits, for instance, move between 3 and 16 kilometers (2 and 10 miles), round trip, in a day. Only under conditions of extreme drought or food shortages do hares move out of their home ranges and roam around in search of food.

Where lagomorphs forage within the home range varies. American pikas, for instance, forage farther from the safety of the talus to collect plants for their hay piles than they do when foraging for immediate consumption. In spring and early summer, European rabbits prefer to forage in areas where grasses and forbs are of intermediate height. Foraging among intermediate-height vegetation may offer some protection against being spotted by predators without compromising the rabbit's ability to detect them. Taller vegetation makes it harder for them to spot predators that may be concealed among the tall vegetation, and shorter vegetation is also easier to handle. However, as the amount of green vegetation decreases with the heat of the summer, the rabbits move to more open areas where green vegetation remains, despite the greater risk of predation.

Biologists have been increasingly interested in the sublethal effects of predators on prey, which may include reduced feeding time due to the need for vigilance, stress (see the discussion of the role stress plays in the snowshoe hare / Canada lynx cycle in the chapter 5 question, "Are rabbits good for the environment?"), and prey foraging in less than optimal food patches if the risk of predation is greater there than in better patches. Ecologist John W. Laundré coined the phrase "the landscape of fear" to describe the amount of risk that prey face from predators in different habitat types. Laundré and his colleagues examined this in black-tailed jackrabbits in the Chihuahua Desert, where coyotes and bobcats are their primary predators. The black-tailed jackrabbits used grassland habitat more than shrubland habitat despite the fact that the grassland has fewer of the forbs, grasses, and cacti they eat, because the open grassland likely enhances their ability to detect an approaching predator compared to the closed shrubland.

A study by Spanish biologist Francisco Palomares and his colleagues measured stress hormone levels in the feces of European rabbits living in Spain's Doñana National Park in four areas of similar habitat but with differing predation pressure. The results showed that, indeed, there was a direct relationship between predation pressure and stress hormone levels, with rabbits suffering the greatest predator pressure being the most stressed. Stress can have a variety of effects in mammals, including reduced immune responses and lower reproductive rates.

As discussed above, lagomorphs do feed very selectively to increase various nutrients and to cope with plant toxins. In an experimental situation, for instance, European rabbits preferred high-protein patches of a grass to low-protein patches of the same species and similar height. America pikas in summer select plants with high nitrogen, which roughly equates to high protein. Snowshoe hares in the summer appear to select plants that offer the highest calorie content (not the highest protein content), modified by the presence of different plant toxins. In winter, they appear to select food items to balance protein and fiber intake, and the twigs they select have higher digestibility than average. Most lagomorphs also tend to select plants with greater water content, especially during dry seasons.

How individual lagomorphs are able to discriminate between plants or plant parts is not very well known but may be primarily a function of taste and smell (see "Why are rabbits always sniffing?" in chapter 2). Black-tailed jackrabbits that live in the desert eat leaves and stems of the creosote bush, which are covered with a bitter-tasting phenolic resin to deter herbivory, with small young leaves and stems more heavily defended than older ones. Jackrabbits appear to nip off both young and old leaves and stems but eat only the young ones, suggesting that taste guides their selection. Similarly, snowshoe hares have been reported to remove the buds of balsam poplar, which are very high in resin, before eating the twigs.

Selection of plants with varying levels of volatile terpenoids, such as those that give sagebrush its characteristic odor, may be mediated by smell. Arctic hares are reported to use their sense of smell to find food buried under snow in winter. Snowshoe hares have also been seen to sniff twigs and then move away from those of species they don't prefer.

The importance of the chemical senses in food selection is supported by experimental studies of the domestic European rabbit. Future food preferences of rabbits are affected by prenatal exposure to the odors of their mother's food. For instance, pregnant rabbits fed thyme produced young that later preferred thyme, while those fed juniper produced young that preferred juniper. This preference may also be established after birth, when young are exposed to the odor of their mother's food in milk and in her fecal pellets.

Do rabbits drink water?

House rabbit owners know that very often their pets drink little water, even though it should always be available to them. The same is true of most lagomorphs. All animals require water, which is necessary for growth,

This creosote bush shows evidence of selective feeding by desert cottontails near the base of the plant, which eat the parts with the least toxins.

Photo © Michael J. Plagens

reproduction (especially lactation), digestion, excretion, and temperature regulation. Water makes up 99% of all molecules within an animal's body and in a 3 kilogram (6.6 pound) domestic rabbit about 58% of the mass is water. Water is constantly being lost during respiration and excretion and must be replaced, but drinking water, or "free water," available in puddles, streams, and other water bodies is not always essential to lagomorphs. Many seem able to get by on "preformed water," the water that is in their food. Also, when food is metabolized into energy, metabolic water is a product of the oxidation that occurs.

We know little about how much free water lagomorphs drink, but the desert-living antelope jackrabbit appears to have no need for free water, and none is available to the island-living Tres Marías cottontail. Where free water is available, it may be more important in sustaining green vegetation than as a source of drinking water for lagomorphs. As discussed above, all lagomorphs prefer succulent vegetation when it is available and this may provide all the water they need in many habitats. Fresh grasses and clover, for instance, may be 80% water; dandelions 85% water, and corn, carrots, broccoli, and lettuce, when rabbits and hares have access to these crops, are 88 to 96% water. During the dry season in the desert, when grasses are gone, antelope jackrabbits eat the succulent pads of cacti to get by. In summer and winter, when green vegetation is abundant, European rabbits prefer to forage where they have the best access to refuge and there is abundant cover, but during the summer dry season they have to abandon areas that provide the best protection from predators in favor of these that provide watery green food.

When succulent forage is not available, lagomorphs have adaptations to conserve water and seem to be able to withstand dehydration better than many other mammals. European rabbits studied in the wild in Australia survived up to 2 months during summer on vegetation containing only 7 to 10% water without drinking free water. During this time, they lost 48% of their body weight in water, showing an ability to cope with dehydration greater than that of a camel, which can tolerate a 40% loss of body weight in water loss, and far greater than dogs, which may die when water loss equals 11 to 14% of body weight.

Some mechanisms that hares and rabbits use to conserve water include concentrating their urine, producing feces with low water content, and behaviorally minimizing water loss due to evaporative cooling by sitting in the shade and reducing activity (see "How do rabbits survive in the desert?" in chapter 5).

Despite such adaptations, black-tailed jackrabbits find it hard to survive under the extremely dry conditions of the Mojave Desert except in years of unusually high rainfall. Here, there appears to be high mortality in summer, and the animals may sometimes be found gathered around water sources. However, frequenting sources of water may increase the risks of predation, which may be why most lagomorphs seem to avoid doing so under most conditions. A study of the use of artificial water sites in Arizona found that lagomorphs—desert cottontails, antelope jackrabbits, and black-tailed jackrabbits—while sometimes seen around the water site, were almost two times more abundant away from these sites. In contrast, predators, including coyotes, foxes, felids, and birds of prey, were more abundant near water sites than away from them.

Some lagomorphs without access to freshwater in water bodies may also take advantage of other sources of free water, such as snow and dew. Arctic and snowshoe hares, for instance, lick snow in winter, when their diet of mostly twigs is quite dry. Mountain cottontails in Oregon scrublands shift their diet regularly, tracking the most succulent vegetation available at the moment. But during the mid-summer drought, they switch to juniper, surprising given it is high in terpenoids and particularly high in that chemical in the morning. It is also lower in protein than some other plants available to them in summer and is not succulent. What's more, these cottontails actually climb juniper trees on summer mornings, a very unusual behavior for a lagomorph, to reach the tips of juniper tree boughs. What junipers do offer the mountain cottontails is dew in the morning, which may be licked from the bough tips or ingested with the soggy foliage.

Hares obtain much of the water they need from their food but may also drink from rivers and other bodies of water. Photo by Andrew Hacking, USFWS

Do rabbits ever store their food?

Rabbits and hares do not store food, but most species of pikas gather vegetation into large hay piles to get them through the winter, when their usual forage of green plants and grasses is scarce and often buried in deep snow in their alpine habitats. The few exceptions, such as some populations of large-eared pikas (*Ochotona macrotis*), are those living where winter snow cover is lacking. The Afghan pika is also unusual in amassing hay piles twice a year: in the spring when plant life is lush for consumption during the hot, dry summer months when plants dry up and again in the rainy fall when plants again flourish. Haying, as it is called, has been well studied in the American pika.

As if predicting the lean times ahead, American pikas amass their winter food stores over about 2 months during the summer, after the end of the breeding season, the period of peak plant abundance. Although they occur simultaneously during the haying season, haying and normal grazing are discrete behaviors. Most obviously, when pikas graze they nibble mostly on short grasses and some forbs and consume the food immediately. When pikas are haying, they nip off relatively tall stalks, leaves, and flowers of forbs and carry their harvest back to a central location, under the talus or tucked in an unobtrusive spot alongside a large boulder. (In contrast, some burrowing pikas make high conspicuous stacks.) American pikas also travel farther to collect hay plants, harvest different plants or plant parts than they eat, and are more selective, with a few species dominating the composition of hay piles compared with the pikas' generally more catholic grazing diet.

American pikas both graze and hay in the meadows surrounding or adjoining the talus fields in which they spend most of their time. The results

The hay pile of an American pika can be quite substantial, weighing 7 kilograms (15 pounds) or more. Photo by Chris Harshaw, Wikimedia Commons / CC-BY-SA 3.0

An American pika carries a long stem of grass to add to its hay pile. This stored food will nourish the pika over the winter. Photo © Andrew Smith

of one careful study showed that grazing pikas stayed as close as possible to the talus, on average venturing no farther than about 2 meters (6.5 feet) into the meadow and, at most, about 10 meters (33 feet) away from safety. To gather plants for their hay piles, however, they moved much farther, on average about 7 meters (23 feet) and as far as 30 meters (98 feet) from the talus. In part, this may be due to the fact that pikas pretty efficiently mow the vegetation closest to their home bases and must go farther afield to find taller plants.

In addition, while American pikas grazing in the summer select plants that are high in nitrogen and low in plant secondary compounds, they actively select plants high in phenolics to store in their winter larder. M. Denise Dearing, who studied this in pikas, offered two interrelated reasons for the behavior. First, plant secondary compounds degrade with storage, so pikas can manage an abundant food source that they could not otherwise take advantage of. Second, the phenolics act as preservatives, so they keep the stored plants from rotting until the phenolics are finally completely degraded.

The hay piles of individual pikas can be quite substantial, weighing as much as 7 kilograms (15 pounds) or more in American pikas, and in some places, present alluring targets for thieves. American pikas, for instance, sometime steal from the hay piles of their neighbors, forcing the pikas to defend them. Other animals sometimes take advantage of pika larders when food is scarce in winter, including reindeer, elk, sheep, hares, voles, and marmots. Where they overlap in distribution, pikas sometimes stockpile marmot scat in their larders—not quite a fair trade. Mongolian herdsmen graze their domestic livestock among Daurian pika populations, where their hay piles remain accessible above the surface of the snow.

Chapter 8

Rabbits and Humans

Do rabbits make good pets?

No wild animal makes a good pet, and lagomorphs are no exception. In fact, lagomorphs in general are challenging to maintain in captivity and are rarely exhibited in zoos. The only lagomorph that makes a good pet is the domestic European rabbit. In the United States, estimates suggest there are about 5 million pet rabbits living in just over 2 million homes, and people who have pet rabbits love them. This number does not include the rabbits raised and shown by hobby breeders, who don't necessarily treat their rabbits as house pets.

But even domestic rabbits are not for everyone, and they are less popular as pets than dogs and cats, which number about 70 and 80 million, respectively, in the United States. Rabbits are more fragile and sensitive than most cats and dogs, for instance. Some experts do not recommend rabbits as pets for families with young children who might inadvertently handle the rabbit roughly, which will, in turn, scratch or bite if it doesn't like the way it's being treated. A houseful of children on the rowdy side is not recommended either because rabbits tend to be stressed by too much commotion. Rabbits can be trained to use a litter box and enjoy socializing and playing with their caretakers but very often do not come when called or may have no desire to sit in your lap. They do chew on everything and your house must be rabbit-proofed so a pet doesn't chew on dangerous items such as electrical wires. Rabbits must also be kept indoors, where they are safe from predators. They must be spayed or neutered to prevent them from marking your home—their home—with urine and feces. The practice of bestowing small children with a bunny rabbit at Easter is generally

The giant Angora is one of several breeds of Angora rabbit, prized for their long, soft fur. Photo by Oldhaus, Wikimedia Commons / CC-BY-SA 3.0

a bad idea—most of these rabbits end up in shelters because families aren't prepared to care for them or can't afford to. If you think a pet rabbit might work for you, there are always plenty of rabbits available for adoption from shelters and rabbit-rescue organizations. The House Rabbit Society advocates for pet rabbits and their owners.

Show rabbits, also called fancy rabbits, are bred by hobbyists to compete for prizes in shows. In the United States, the American Rabbit Breeders Association registers and sanctions rabbit shows, breeds, and organizations. The ARBA recognizes 45 rabbit pure breeds, the most popular of which are Mini Rex, Netherland Dwarf, and Holland Lop, all relatively small breeds that are also popular as pets. The European Association of Rabbit Breeders, which includes 15 countries, registers 66 breeds. Rabbit breeds differ primarily in size, coat color and length, carriage, and length of the ears. Breeding rabbits for show began in the late 1800s, about the same time as people began breeding dogs and cats for the same purpose. The domestication of rabbits has a longer history.

Whether you are interested in adopting a rabbit as a house pet or in breeding rabbits for show or as a commercial activity, there are many excellent guides to their care. The Web sites of the House Rabbit Society (www.rabbit.org) and the American Rabbit Breeders Association (www.arba.net) offer good introductions. *Rabbit Production* by Peter R. Cheeke, Nephi M. Patton, Steven D. Lukefahr, and J. I. McNitt, now in its eighth edition, is considered the bible of rabbit care, primarily as it relates to breeding rabbits commercially. *The House Rabbit Handbook: How to Live with an Urban*

The silver is one of many breeds of domestic rabbit. Photo by Meghan Murphy, Smithsonian's National Zoo

Rabbit by Marinell Harriman, now in its fourth edition, is a highly regarded primer on care for pet rabbits.

How were rabbits domesticated?

More than 2,000 years ago, ancient Romans began systematically exporting European rabbits from Iberia to the Mediterranean islands and Italy and eventually to all parts of the empire. The rabbit is believed to have reached even as far as China during that period as a result of the Roman's vast trade network along the Silk Road.

Romans kept rabbits, hares, and often deer and birds in leporia, enclosed spaces several acres in size. Sometimes a small island served as a leporia without the attendant cost of building a walled enclosure. Rabbits were hunted in the leporia, often with ferrets, which were muzzled and sent to chase the rabbits out of their burrows so they could be caught by people wielding nets. In fact, this is why ferrets themselves were domesticated.

The Romans did not deliberately domesticate rabbits, however. They simply caught them, held them, and ate them. Many rabbits escaped from the leporia—not so hard given the animals' ability to burrow under and out—and these were likely the least tame individuals. Most escapees were probably taken by predators, but a few survived to begin wild populations far from their native haunts. As people transformed the European landscape from forest to farm, the decline of predators further abetted the spread of rabbits and today they live in most of Western Europe.

Medieval French monks are credited with domesticating rabbits between 500 and 1000 CE. The monks had very special needs for the rabbit as food, apart from a taste for rabbit meat. At a time in Catholic Europe when even lay people had more fast days on the calendar than not, fetal or newborn rabbits—long a delicacy—were deemed "fish" by the Church authorities. Thus, monks could dine quite well on "laurices," as the fetal rabbits were called. For the monks to monitor pregnant females and be on hand at the moment of birth, the laissez-faire Roman style of keeping rabbits had to go. The monks instead kept their rabbits in smaller high-walled and paved courtyards, forcing the rabbits to breed and give birth on the ground.

It is hard to imagine that at one time rabbits were considered rare animals, prized for their meat, fur, and hides, with tens of thousands of rabbit skins being exported each year during parts of the Middle Ages. Because of the high value of rabbits, controlling and maintaining warrens became the province of the king and the nobility, all of whom hired warreners to look after and cull rabbits. After about 1100, with the decline in Western Europe of game animals and fur-bearing animals such as bears and beavers, landowners in Great Britain and France began setting aside large pieces of land for rabbit warrens, which they carefully managed. Predators were eliminated, food was supplemented, and holes in the soil were even dug to facilitate burrowing.

Even in these situations, though, the rabbits were basically wild animals given a bit of extra help from people so they could be hunted. It was here, however, that selection for larger size began; smaller females and old males were hunted and larger females were left in the warrens to breed again. This likely explains why the wild rabbits of northern Europe, which are descendents of feral warren rabbits, are larger today than those in the south. Monks continued to breed rabbits in this period; there is even a story about an abbot giving King Edward III 60 rabbits as a gift.

Gradually, hutches to contain the rabbits were introduced, and, by about 1600, breeders were selecting for coat color, with gray, silver gray, black, and white rabbits on offer to provide variety for their fur-seeking customers. The Champagne d'Argents, a silver rabbit from France, is known from the mid-1500s. Lop-eared rabbits and Angoras were relatively early breeds too, appearing in the 1700s. The Belgian "hare," a rabbit breed with a harelike shape, was not developed until the mid-1800s and many other breeds only came into being in the twentieth century. In addition to the breeds defined by standards set by the formal rabbit breeding associations, commercial strains bred for certain production traits were developed fairly recently, and local populations of backyard rabbits have evolved different traits in different places. Local populations have been poorly described and are disappearing.

Around the beginning of the 1800s, the right to maintain rabbits in leporia and warrens, indeed the right to hunt rabbits at all, spread beyond royalty and wealthy landowners in Great Britain and some other parts of Europe, so both rural and urban people began raising rabbits in backyard hutches for food, a practice that also appeared about the same time in the United States, where hunting cottontails and hares for the plate has a long history. Taking care of these hutch rabbits was generally considered women's work, because hutches were close to home and the rabbits were small and tractable. This remains the case today in many parts of the developing world where rabbit breeding has been introduced to improve human diets and livelihoods.

All domestic mammals exhibit a suite of traits that differentiate them from their wild ancestors and surprisingly, are similar across a range of species from rabbits, dogs, rats, and cats to sheep, goats, and pigs. These traits include the appearance of dwarf and giant breeds, piebald coats with splotches of different colors, wavy or curly hair, floppy ears, and changes in reproduction to permit year-round breeding. Domestic mammals, including rabbits, also have smaller brains and generally poorer vision and hearing.

A fascinating experiment conducted on foxes suggests why this might be so. Beginning in 1959, Dmitry Belyaev bred a line of red foxes, starting with wild individuals, and subjected them to strong selection for only one trait: tameness. Within a mere 20 years, Belyaev had produced a large number of unaggressive foxes, some of which were piebald and had floppy ears and shortened snouts, and some females had started to breed twice a year. This suggests that whatever is responsible for "tameness" in mammals is involved in a host of other physiological processes, and some scientists have suggested there is a domestication gene or set of genes operating in all mammals.

Some evidence indicates that the physiological and behavioral changes associated with domestication are mediated by the endocrine system, perhaps via thyroid hormones, which affect every cell in the body. Domestic rats, for instance, have smaller thyroid glands than wild rats. Other evidence suggests a role for the neurotransmitter serotonin, which modulates aggression and mood and has a host of other physiological effects. One example were Belyaev's tamed foxes, which had elevated serotonin levels, as do domestic rats. In people, increasing serotonin levels reduces anxiety and depression. Moreover, serotonin is converted to melatonin in the brain, a hormone involved in the regulation of activities and processes governed by photoperiod. These include seasonal reproductive activity and seasonal changes in fur growth and color. Melatonin also has other far-reaching physiological effects.

Recent genetic analyses show that domestic rabbits have relatively low levels of genetic diversity, compared with the wild rabbits of France, which data suggest are the progenitors of most domestic rabbit breeds. France's wild rabbits, in turn, have less genetic diversity than those of the Iberian Peninsula, where genetic diversity is high. As noted earlier, the rabbits of the Iberian Peninsula are divided into two evolutionary divergent populations, one in the southwest and one in the northeast, and all of the rest of the world's wild and domestic rabbits are descended from northeast population. European hares are bred in captivity in parts of Europe but they are very difficult to maintain and they experience stress, high levels of disease and infant mortality, and reproductive problems.

Why did people say "the rabbit died" to mean a woman was pregnant?

In the middle of the last century, laboratory domestic rabbits played a role in diagnosing human pregnancy. It worked like this. Urine from a woman was injected into a vein in a female rabbit's ear, and 2 days later the rabbit was killed so its ovaries could be examined. If the woman was pregnant, the rabbit would have ovulated by then, in response to a hormone, called human chorionic gonadotropin (hCG), that is excreted in the urine of pregnant women. If the rabbit had not ovulated, there was no pregnancy. The rabbit died in any case—for no other reason than because it was killed so the ovaries could be seen—although killing the rabbit was expedient but not essential.

In an episode of the long-running television show *M*A*S*H*, set in the 1950s Korean War, Margaret "Hot Lips" Houlihan feared she was pregnant and turned to Dr. Hawkeye Pierce for help. The only rabbit available was Corporal Radar O'Reilly's pet, and he was understandably reluctant to sacrifice Fluffy. So Pierce performed careful surgery to check out Fluffy's ovaries, sewed up the incision, and the rabbit was shown recovering under the relieved Radar's care (*M*A*S*H*, Episode 142, "What's Up, Doc?" January 30, 1978).

It's not known how "the rabbit died" became a euphemism for pregnancy given that, except on television, the rabbit *always* died. Perhaps it has to do with it being unusual and unacceptable for women at the time to widely disseminate news on a nonpregnancy, so announcing she was tested was enough to mean she was pregnant.

Rabbits were not the first or only animals to be used as bioassays of pregnancy. An earlier test involved injecting urine into an immature female rat or mouse, which would subsequently go into estrus if hCG was present in the urine. African clawed frogs were also used, which would produce eggs if injected with hCG-laden urine. Like the rabbit test, these were time-consuming, expensive, and not always accurate; for instance, luteinizing hormone, which is released by the pituitary gland and in females controls many facets of reproduction, could have similar effects.

In the 1960s, as research on hormones expanded and interest in better diagnosing pregnancy grew, these crude bioassays were replaced with the first crude immunoassays, in which antibodies to hCG, developed by immunizing rabbits, were added to urine. Antibodies to hCG were harvested from rabbits and, when the antibodies were mixed with a pregnant woman's urine, produced a particular pattern of red blood cell clumping in the sample, but this test was not without problems either, including cross-reaction with luteinizing hormone. The real breakthrough came in the early 1970s, when a researcher at the National Institutes of Health was able to generate antibodies, again in rabbits, to the specific portion of the hCG molecule that was responsible for its biological activity. This meant the antibodies did not cross-react with other hormones and that radioactive labels could be attached to them for the assay. By 1978, the first over-the-counter home pregnancy test kits were on the market, and today, a few drops of urine and a kit provide instant results very soon after fertilization. Antibodies from rabbits are sometimes used in these kits, along with antibodies from mice.

Are rabbits used in a lot of experiments?

Domestic rabbits have been used as subjects in scientific experiments at least since the early 1800s, when a German scientist studying rabies, Georg Gottfried Zinke, experimentally infected healthy rabbits with the disease by brushing saliva from a rabid dogs into an incision in the rabbit. Later in that century, rabbits became the experimental model of choice in rabies studies, perhaps because infected rabbits usually exhibit a form of the disease that causes paralysis rather than the "furious" behavior of dogs. In 1885, French scientist Louis Pasteur used rabbits to develop the first successful vaccine to treat rabies in a person. Pasteur injected live rabbits with rabies virus and then killed the infected rabbits and removed and dried their spinal cords. The treatment consisted of injecting the human victim with the emulsified spinal cord over 10 days, starting with less virulent cords that had been dried for 2 weeks and progressing to more virulent ones dried for less time.

Even earlier than that, though, seventeenth-century Dutch scientist Regnier de Graf first identified the structure on a rabbit's ovary that came to be known as a Graffian follicle, which is the mature stage of an ovarian follicle that ruptures to release an egg capable of being fertilized.

Another scientific milestone involving rabbits was the first embryo transfer in 1890, when two Angora rabbit embryos were implanted in a pregnant Belgian hare (a breed of domestic rabbit). The scientist was interested in the effect of the uterine environment on the phenotype (external appearance) and his results revealed its negligible effect—the rabbit gave birth to a mixed litter of Angoras and Belgians.

Domestic rabbits possess several traits that have made them prominent in laboratory research. They are easy to obtain, with many bred for the sole purpose of supplying research animals. The New Zealand white rabbit, which contrary to its name was developed in the United States for meat and fur production, is the most frequently used breed. Domestic rabbits are docile, so they are easy to handle, and prominent veins in the ears make it simple to draw blood. Short gestation periods are an advantage in studies related to reproduction; rabbits are often used to test whether chemical exposure results in abnormalities to developing fetuses, for example.

Rabbits also display, or can be induced to display, some of the same diseases and genetic abnormalities that people do, including diabetes, tuberculosis, cancer, and heart disease, making them good model organisms. The first link between dietary fat and the incidence of atherosclerosis (the narrowing and hardening of arteries that can lead to strokes and heart attacks) was discovered in a 1908 study of rabbits fed diets of eggs, milk, and meat.

In 1998, Robert F. Furchgott, Louis J. Ignarro, and Ferid Murad won the Nobel Prize in Physiology or Medicine for "Nitric Oxide as a Signaling Molecule in the Cardiovascular System," a study that was conducted on rabbits. Nitric oxide is an atmospheric pollutant produced when nitrogen is burned and present, for example, in car exhaust. But the Nobel Prize–winning scientists found that it is also a naturally occurring neurotransmitter that tells blood vessels to dilate or contract. Nitric oxide has since been shown to play a role in cardiovascular disease, as well as in immune responsiveness and memory formation. Nitroglycerine, which is often prescribed to reduce the pain of angina due to restricted blood flow to the heart, does so by generating nitric oxide, which relaxes the walls of the coronary arteries. Nitric oxide is also the key ingredient in Viagra.

Although research on live animals such as rabbits is controversial, there can be no doubt that scientific studies using rabbits as subjects have considerably advanced our understanding of diverse aspects of human biology.

Many rabbits are also used to test the safety of various products from cosmetics to pesticides before they are approved for use on or around people. The albino New Zealand whites, for instance, are used for the Draize

Eye Irritancy Test. The Draize test analyzes the effects of chemicals on the eyes, and the unpigmented eyes of these rabbits make it easy to visualize any pathological changes resulting from chemical application. Justifiably criticized as inhumane, and by many as unreliable, the use of this test is slowly being phased out as other tests that do not involve the use of live rabbits, such as testing on cell cultures, are developed.

The number of rabbits used in research has declined significantly in recent years, from about 550,000 per year in 1987 to about 200,000 per year today. There is a growing scientific effort in the United States and in Europe to find alternatives to using live animals in research and testing and, where scientists must use animals, to minimize the numbers involved. The Animal Welfare Act of 1966 improved the standard of care given to laboratory animals, including rabbits (but excluding rats and mice, which are used in research vastly more often than rabbits or any other species). For instance, every effort must be made to minimize or alleviate the pain caused by any research procedure (see "Do rabbits feel pain?" below).

In a recent development, scientists are using pikas to study leptin, a hormone that is produced in brown adipose tissue (brown fat; see "How do rabbits survive the winter?" in chapter 5). Leptin is involved in regulating appetite and metabolism, with high levels of leptin reducing appetite and increasing energy expenditure, and low levels the reverse. Animals that are losing fat, for instance, produced less leptin, telling the body to eat more and spend less. Abnormalities in responsiveness to leptin are linked to obesity and associated diabetes in people. Unlike many mammals, pikas do not appear to eat less, lose weight, or decrease their metabolic rate in response to the very cold temperatures in which they spend the winter. They also show no seasonal changes in leptin production. Scientists report that there appears to be positive selection on genes coding for leptin in pikas, with some unique genome sequences not seen in other mammals or lagomorphs. They believe a better understanding of leptin in pikas might contribute to new insights into human metabolic disorders related to this neurotransmitter.

A study of lagomorph genetics may provide insight into the evolution of the kind of viruses that cause AIDS in people and similar diseases in nonhuman primates and cats. HIV, which causes human AIDS, is a retrovirus. Retroviruses have the insidious ability to worm their way into the genomes of mammals—including yours—and then are passed down to offspring. Some 5 to 8% of our genome is composed of remnants of retroviruses that were once infectious. Until recently, however, HIV and related retroviruses, called lentiviruses, had not been found in the genomes of any mammal. Instead, they were known only to jump between species—HIV is believed to have evolved from a lentivirus that jumped from a great ape

New England cottontail (*Sylvilagus transitionalis*). Photo by Linda Cullivan, USFWS

Omilteme rabbit (*Sylvilagus insonus*) is known only from a few skins.

Photo © Susan Lumpkin

Tapeti (*Sylvilagus brasiliensis*).

Photo © Mario Sacramento

Snowshoe hare (*Lepus americanus*). Photo by Walter Siegmund (with permission), Wikimedia Commons / CC-BY-SA 3.0

Black-tailed jackrabbit (*Lepus californicus*). Photo by Scott Rheam, USFWS

White-tailed jackrabbit (*Lepus townsendi*). Photo © Jim Witham

Tehuantepec jackrabbit (*Lepus flavigularis*) young. Photo ©
Tamara Rioja Paradela

European hare (*Lepus europaeus*).
Photo © Silviu Petrovan

Arctic hare (*Lepus arcticus*).
Photo from USFWS

Irish mountain hare (*Lepus timidus hibernicus*). Photo by Alan Wolfe, Wikimedia Commons / CC-BY-A 3.0

Broom hare (*Lepus castroviejoi*).

Photo by Jorgenix, Wikimedia Commons / PD

Cape hare (*Lepus capensis arabicus*).

Photo © Shah Jahan

Woolly hare (*Lepus oiostolus*).

Scrub hare (*Lepus saxatilis*).

Indian hare (*Lepus nigricollis*). Photo © Gehan de Silva Wijeyeratne

Ethiopian highland hare (*Lepus starcki*). Photo by Jeff Kerby, Wikimedia Commons / CC-BY-A 2.0

Research on pikas may shed light on human metabolic disorders. These lagomorphs have unique genetic coding of the hormone leptin that regulates appetite and metabolism.

Photo © Andrew Smith

to people. This long puzzled scientists studying them. In 2007, a team of scientists discovered that the European rabbit genome carries a lentivirus fragment, which they named RELIK. In 2009, another team found similar lentivirus fragments in the genomes of Granada hares, tapeti, and riverine rabbits, but not pikas. Thus lagomorph ancestors appeared to have acquired the lentivirus after the split between the pika and leporid lineages but before rabbits and hares split about 12 million years ago (see "When did rabbits evolve?" in chapter 1).

Rabbits also figure in a famous thought experiment by Leonardo Fibonacci, a medieval Italian mathematician. Fibonacci posed the following problem: Put a single pair of rabbits in a confined place where the pair gives birth to another pair at the end of the second month. Each resulting pair produces a second pair from the second month on. How many pairs of rabbits will there be at the end of each month in a year? The answer is the number sequence 1, 1, 2, 3, 5, 8, 13, 21, 34, 55, and so one, in which each number is the sum of the two preceding numbers. This was the first example in Europe of what mathematicians call a recursive number sequence. Later mathematicians discovered that as the numbers grow greater, the ratio between succeeding numbers nears the golden ratio (the number 1.6180 . . .), which extends forever without repeating the same sequence of digits. The Fibonacci numbers, also called the Fibonacci sequence, have been found to describe many natural phenomena, such spiraling petals of a sunflower and the whorls of pine cones, leaf buds on a stem, and the curve of animal horns.

Do rabbits feel pain?

Of course lagomorphs feel pain, just as all mammals do because their nervous systems are all very similar. Wild animals are generally adapted to attempt to hide pain in order not to call attention to themselves from predators or competitors. Physical pain is generally the result of tissue injury, which causes changes in the associated cells. These changes release biochemicals that activate specialized nerve cell endings called nociceptors. With enough stimulation, nociceptors fire and signals move via nerve fibers to the spinal cord and the brain, where the experience of pain is generated.

Domestic laboratory rabbits are often used to test the efficacy of analgesics (painkillers), for instance by administering the analgesic and determining the test animal's threshold for responding to measured levels of pressure or heat. A common method is to apply the stimulus to the bottom of a rabbit's foot and see how long it takes for the rabbit to withdraw it. Studies have also been performed on rabbits to investigate the mechanisms underlying the analgesic effects of acupuncture.

Increased concern over animal welfare, especially regarding laboratory animals, and its regulation by governmental agencies under the U.S. Animal Welfare Act and other legislation, has led scientists to examine the issue of pain in animals in some detail. The American College of Laboratory Animal Medicine recently published a position paper called "Guidelines for the Assessment and Management of Pain in Rodents and Rabbits." This paper defines pain quite broadly, as "an unpleasant sensory and emotional experience associated with actual or potential tissue damage, and should be expected in an animal subjected to any procedure or disease model that would be likely to cause pain in a human." Among the signs listed as signifying that a rabbit is in pain are reduced activity, not grooming, eating and drinking less than usual, eye squinting, hunched posture, hiding or aggressiveness, and teeth grinding. Recommendations for preventing, minimizing, and alleviating pain are very similar to those recommended for humans, ranging from use of anesthetics and analgesics to keeping the animal warm, dry, and comfortable.

What if I find a baby rabbit or an injured rabbit or if I hit one with my car?

If you find a baby rabbit or hare, or a nest of babies, in all likelihood they are just fine. We earlier described the absentee maternal care system of rabbits and hares—moms visit babies for only a few minutes a day to nurse them and the rest of the time stay away from them. Young cottontails

A baby cottontail on its own has most likely not been abandoned by its mother. It is best to leave it alone.
Photo by Jim Peaco, NPS Photos

also leave the nest and are weaned at 3 to 4 weeks of age, hares sooner still, so it is normal to see young rabbits on their own. Thus, seeing a baby alone does not mean it has been abandoned. The best thing you can do is simply go away and leave the baby or nest alone. If a cottontail nest has clearly been disturbed and small babies are displaced from it, you can gently move the babies back into the nest, then go away—mom will not return to the babies if you are in the way. Baby hares or jackrabbits, which are precocial and freeze when approached, should not be handled. In attempting to escape from your hands, they may injure or kill themselves. Even with your good intentions, you are just another potential predator to these animals.

If you are sure that babies have lost their mother—for instance, if your dog or cat has killed the mother or if the babies are very cold or crying, suggesting they are not being cared for—you should contact a veterinarian, your animal control department or local humane society, or an animal shelter for assistance and referral to a licensed wildlife rehabilitator. It is extremely difficult to raise orphaned rabbits; according to the House Rabbit Society, fewer than 10% of orphaned rabbits survive more than a week. In many states, it is also illegal for anyone other than a licensed rehabilitator to care for a wild animal.

The Web site of the House Rabbit Society offers an excellent overview of what to do in these situations. Or, in the case of orphaned or injured hares and rabbits, you can simply let nature take its course. As hard as it may be to walk away from a suffering animal, it will very quickly provide a meal for a hungry predator.

If you find an injured rabbit or hare or have hit one with your car and it is alive but injured, you should not attempt to care for it yourself. Again, the best thing to do is call your local animal control, humane society, or animal shelter for advice and assistance.

The official joke of the Lagomorph Specialist Group addresses hitting a rabbit with your car:

> A man was driving down the highway, and he saw a rabbit hopping across the road. He swerved to avoid hitting the rabbit, but unfortunately the rabbit jumped in front of the car and was hit.
>
> The driver, being a sensitive man as well as an animal lover, pulled over to the side of the road, and got out to see what had become of the rabbit. Much to his dismay, it was dead. The driver felt so awful, he began to cry.
>
> A woman driving down the same road came along, saw the man crying on the side of the road, and pulled over. She stepped out of her car and asked the man what was wrong.
>
> "I feel terrible," he explained. "I accidentally hit this rabbit and killed it."
>
> The woman told the man not to worry; she knew what to do. She went to her car trunk, and pulled out a spray can. She walked over to the limp, dead rabbit, and sprayed the contents of the can onto the animal.
>
> Miraculously the rabbit came to life, jumped up, waved its paw at the two humans, and hopped down the road. Fifty yards away, the rabbit stopped, turned around, waved again, hopped down the road another fifty yards, waved and hopped another fifty yards.
>
> The man was astonished. He couldn't figure out what substance could be in the woman's spray can!! He ran over to the woman and demanded, "What was in your can? What did you spray on that rabbit?"
>
> The woman turned the can around so that the man could read the label. It said: "Hair spray. Restores life to dead hair. Adds permanent wave."

How can I see rabbits in the wild?

The first thing to do is find out what species might live in your area, using this book and other reference materials. If you live in an urban area, you may need to visit a park or other natural area for a chance to see a lagomorph. Most national parks in the United States and Canada have websites that list the species that can be found in them, as do many state and provincial parks and some nature centers. The Web sites of state and provincial fish and wildlife departments will also be helpful. Rural residents in most parts of North America will have little trouble seeing the rabbits and hares that live in their area. Walking or driving slowly along a country road at dawn or dusk, you may see a bunny foraging in a clearing between the road and the woods, for instance.

Following its tracks in snow is a good way to find a jackrabbit.

Photo © Jim Witham

Once you urbanites have found a place you're likely to see a lagomorph, determine what particular habitat types it lives in to narrow your search in the field. When visiting a park, you can also ask a ranger if he or she knows a likely place to spot one. Lagomorphs in general don't move over very large areas, so if a ranger has seen one recently, it's very likely a rabbit or hare will still be nearby; in the case of pikas, they will almost certainly be in a patch of talus. You can see pikas during the day, but the best time to look for rabbits and hares is dawn and dusk. Once you settled on a potentially rewarding bunny-watching site, sit quietly and be patient. And should a rabbit or hare come into sight, don't move at all—unless you want to see how fast a lagomorph can run!

If there is fresh snow on the ground, you can walk slowly through rabbit or hare habitat and look for tracks in the snow. Following the tracks may lead you to its shelter where it may be visible.

If you spot a bunny, observe its physical traits and behavior and note where it is, the time of day, and other observations. Sketch it or take a picture or video if you have a camera. If more than one species lives in your area, you may want to take a field guide to identify which one you're seeing, or compare your photo to pictures in this book or on a Web site when you get home.

Wildlife watching is very rewarding and some people make a career of it, as a scientist or nature photographer, for instance. But it's also a great hobby. Even if your walk in search of bunnies doesn't reward you with a sighting every time, you can enjoy watching birds, identifying trees and wildflowers, or just breathing the fresh air.

Should people feed rabbits?

In the normal course of things, there is no reason for people to feed any wild rabbit, hare, or pika. It is better to let these animals find and select their own food as they are adapted to choose the most nutritious vegetation available in their natural habitat.

Chapter 9

Rabbit Problems (from a human viewpoint)

Are rabbits pests?

One of the first historical mentions of European rabbits is from the Greek geographer Strabo, who lived from about 58 BCE to 20 CE. He, at least, saw them as pests, writing,

> Turdetania [modern-day Andalucía province in southern Spain] also has a great abundance of cattle of all kinds, and of game. But there are scarcely any destructive animals, except the burrowing hares, by some called "peelers"; for they damage both plants and seeds by eating the roots. This pest occurs throughout almost the whole of Iberia, and extends even as far as Massilia [Marseilles, France], and infests the islands as well. The inhabitants of the Gymnesiac Islands [the Balearic Islands of Majorca, Menorca, and others in the Mediterranean off the coast of eastern Spain], it is said, once sent an embassy to Rome to ask for a new place of abode, for they were being driven out by these animals, because they could not hold out against them on account of their great numbers. Now perhaps such a remedy is needed against so great a warfare (which is not always the case, but only when there is some destructive plague like that of snakes or field-mice), but, against the moderate pest, several methods of hunting have been discovered; more than that, they make a point of breeding Libyan ferrets, which they muzzle and send into the holes.

It is worth noting that rabbits were not native to the Balearic Islands but had been introduced there as early as about 1350 BCE by settlers from the Iberian Peninsula. These islands had no native mammalian predators to control rabbits either.

Writing about a century later, Pliny the Elder, a Roman natural historian, tells a similar story about rabbits in the Balearic Islands. In his account, the rabbits reportedly ate so much grain that the people were starving, so the besieged residents petitioned the Roman emperor to send troops to kill the rabbits or at least cart them away. Pliny adds elsewhere that the city of Tarragona in Spain was completely destroyed by rabbits. That is very likely an exaggeration, but in the early twentieth century, a lighthouse keeper released rabbits on Washington's San Juan Island, where 20 years later their tunnels threatened the lighthouse with collapse.

More recently, archeologists in Great Britain are struggling with rabbits undermining ancient sites. The remains of Hadrian's Wall, for instance, which was built across northern England by Romans in 122 CE to forestall the invasion of ancient Scots, is at risk. Quoted in the UK newspaper *The Guardian*, a Liverpool University archeologist said: "Some [sites] look like the surface of the moon. Rabbit burrows have created such a honeycomb beneath sites that sooner or later there will be a single catastrophic incident where the whole thing vanishes."

In his masterful book, *Ecological Imperialism*, Alfred W. Crosby writes of the first Portuguese who settled the island of Porto Santo, which lies east of North Africa near Madeira in the Atlantic, in the 1420s. They released a single female rabbit (and the young she delivered on the ship) on this previously uninhabited island, with no evidence of any earlier human presence. In the absence of any predators or diseases to which the animals were susceptible, these rabbits bred so prolifically, at a "villainous rate," and ate so rapaciously that the settlers' attempts to grow crops failed completely. Eventually, the settlers had to abandon the island, "defeated in their initial attempts at colonization not by primeval nature but by their own ecological ignorance."

Apart from their impact on human affairs, introduced European rabbits have wreaked ecological havoc around the world. People have introduced both domestic and wild European rabbits on some 800 islands or island groups. The reasons for these vary. In addition to those liberated to provide food for seafarers, rabbits were released for sport hunting, to raise for food or fur, as food for other animals (even as lobster pot bait), and to control vegetation.

Hawaii's Laysan Island is a notorious example, but just one of many, of the effects of ill-advised rabbit introductions. Laysan lies about 1,450 kilometers (900 miles) north of Oahu. The following is based on a review by ornithologist Storrs L. Olson. On its discovery by whalers in about 1820, it had no human inhabitants, but its bird life was spectacular. Along with 17 species of breeding water birds, it boasted five land birds found nowhere else. In 1890, guano (bird dung) mining came to the island, greatly

The burrows dug by overabundant European rabbits may lead to erosion and landslides and sometimes undermine archeological ruins. Photo by Brammers, Wikimedia Commons / PD

disturbing the small island's ecology. In the early 1900s, Laysan also attracted gangs of Japanese feather collectors, who killed thousands of birds to export tons of feathers used to trim hats. But most devastating of all, Max Schlemmer, self-styled emperor of Laysan and superintendent of the guano-mining operation, released rabbits to the island in about 1903, hoping to start a meat-canning factory there. As elsewhere, the rabbits bred like rabbits and the island's vegetation began to suffer. Concern over this, combined with feather poaching, led then-president Theodore Roosevelt to declare the island and others in the chain the Northwestern Hawaiian Islands Bird Reservation, but that didn't stop the multiplying rabbits. In 1911, there was a call to eradicate the rabbits, and the next year some 5,000 rabbits were shot. But because not every last one was killed—the hunters ran out of ammunition—the rabbits continued to eat away.

Finally, in 1923, a joint expedition of the U.S. National Biology Survey (forerunner of the U.S. Fish and Wildlife Service), the U.S. Navy, and Hawaii's Bishop Museum was sent to explore the biological status of the Northwestern Islands (now part of the Papahanaumokuakea Marine National Monument created by in 2006). In particular, they were charged with eliminating the rabbits once and for all from Laysan Island. The Tanager Expedition was led by Alexander Wetmore, who would later go on to be secretary of the Smithsonian Institution. Ironically, among the expedition members was the son of Max Schlemmer.

What they found on Laysan was a barren wasteland, so devoid of green vegetation that even the rabbit numbers were reduced—only few hundred or so had to be shot or poisoned. But it was too late for two of the land birds—the Laysan millerbird (*Acrocephalus familiaris familiaris*), an Old

World warbler, and the Laysan 'apapane (*Himatione sanguine freethii*), a honeycreeper. They were already extinct when the Tanager Expedition arrived. The last three individuals of a third species, the Laysan rail (*Porzana palmeri*), died in a sandstorm while the expedition team was on the island. Two others, the Laysan duck (*Anas laysanensis*) and Laysan finch (*Telespiza cantans*), were reduced to perilously low numbers and both remain endangered today. Twenty-six plant species had also been eradicated by the rabbits. The 'apapane perished because this nectar-eating birds' food disappeared into the rabbits. The millerbirds starved to death as the moths on which they depended died when the rabbits devoured their food plants. In fact, six of the eight species of moths on the island also went extinct.

Sometimes the deleterious effects of rabbits are more indirect. The extinction of a parakeet (*Cyanoramphus erythrotis*), native to Australia's subantarctic Macquarie Island, is a case in point. Discovered by sealers in 1810, treeless Macquarie Island, which lies about midway between New Zealand and Antarctica, was home to the abundant ground-nesting parakeet along with immense numbers of seabirds. For the next 70 years, the parakeets continued to do fine, despite their being hunted with dogs and eaten for food by sealers. They also withstood the introduced cats and wekas (*Gallirallus australis*)—predatory birds native to New Zealand, likely because neither of these predator became very abundant. They were limited by food scarcity, especially in winter when few seabirds remained on an island with no small mammals and only the parakeet and a rail as resident land birds. Until, that is, rabbits were released there in 1879 and by 1884 were counted by the thousands. With a year-round food supply, cat and weka numbers skyrocketed by 1884. By 1890, just 6 years later, the parakeets were gone, falling victim to the cats and wekas. This likely also accounts for the banded rail's (*G. philippensis macquariensis*) extinction a few years later.

Today, rabbits are still a problem on Macquarie Island, which is a World Heritage Area and a UNESCO Biosphere Reserve. The rabbit population reached a peak number of about 150,000 in the 1970s. The *Myxomatosis* virus was introduced to control them about the same time, and numbers subsequently fell. But with successful efforts to eradicate both the wekas and the cats, completed by 2000, and the rabbits' growing resistance to the myxoma virus, their numbers surged again. Warmer and drier winters associated with climate change have also improved rabbit breeding success. As a result, the island's vegetation is severely overgrazed leading to erosion and slope instability. This is destroying nesting habitat for birds, especially of burrow-nesting petrels but also of others. With the vegetation that once helped conceal nests and chicks gone, predatory skuas—whose numbers are inflated by a rich fare of rabbits—are having a field day.

In 2007, landslides buried a large number of king penguins (*Aptenodytes patagonicus*) and their nests. At least 24 bird species, including several species of petrels, prions, and albatrosses, are suffering and several no longer breed on the island, having taken refuge on offshore rocks. The presence of black rats and house mice, which eat seeds as well as chicks and eggs, exacerbates the problem. What's more, rabbit overgrazing of native vegetation, such as unusual megaherbs, gives a boost to introduced plant species, which keep native vegetation from recovering. This, in turn, is affecting the about 350 species of invertebrates that live on the plants. In short, this is a mess, so much so that in 2007 an eradication plan was developed that aims to eliminate the rabbits and rodents once and for all.

With current technology and sufficient resources, it is possible to eradicate rabbits from relatively small islands. It's another story, however, on larger landmasses, most famously in Australia.

The British First Fleet, sent to begin the first European colony in Australia, arrived in what is now Sydney in 1788 with five domestic rabbits on board. Subsequently many more were released, largely to the efforts of "acclimatization societies," which existed basically to make Australia more like home, that is, more like England (where, ironically, European rabbits were no more native than in Australia), where rabbit hunting was as important as sport as for putting meat on the table.

Coming from domestic stock, none of these introductions really took off. In 1859, however, a man named Thomas Austin released 20 wild rabbits from England in Victoria. More wild rabbits were released in South Australia about 1870. From this handful of individuals, rabbits in enormous numbers soon marched across vast swathes of southern Australia like a hungry barbarian army. If the rabbits didn't exactly sack, they certainly did pillage. Farmers and ranchers were helpless to protect their crops and pastures from these little eating machines. By 1879, rabbits had so thoroughly infested western Victoria (which abuts South Australia), that farmers were driven off their land. From this starting point, the rabbits marched inexorably across the continent, in the most favorable habitats at the rate of 125 kilometers (77 miles) a year and along some river systems as fast as 300 kilometers (186 miles) a year. Despite farmers and ranchers poisoning, trapping, and shooting rabbits and the introduction of predatory foxes and cats, rabbit numbers grew and grew. Just 30 years later, the government was offering a 25,000-pound reward (a fortune in today's dollars) to anyone who could devise a solution to the rabbit plague. Within 60 years, rabbits had conquered essentially the entire continent but for the far north. In contrast, it took about 700 years for rabbits to fully colonize Great Britain.

Even Australia's infamous rabbit-proof fences—the country's counterpart to China's Great Wall—failed to block the rabbit's invasion. Built in

Western Australia between 1901 and 1907 to keep rabbits out of outback pasturelands, the rabbit-proof fences stretched about 3,200 km (2,000 miles). One fence (No. 1) went north and south from coast to coast, with another (No. 2) running parallel along much of it about 130 km (80 miles) apart, the rabbits having breached the first fence before it was complete. Another fence, which connected these two from east to west, was known as No. 3. Difficult and expensive to build through the rough country and even more so to maintain through fires, floods, and sandstorms, the fence never really worked very well to keep rabbits out, but it did keep kangaroos and emu out of the grazing lands. And it is interesting that Australian scientists have noticed a marked difference between rainfall on either side of the fence. There is about 20% less rainfall on the agricultural lands on western side of north-south fences than on the eastern side where natural vegetation reigns.

Rabbits found near-ideal conditions in their adopted land, some natural, some human-induced. The warm, dry climate of most of the southern half of Australia is similar to that of the Mediterranean, the rabbit's natural home. The burrows of several native marsupials, such as wombats, rat-kangaroos, and the bilby (*Macrotis lagotis*), provided ready-made shelter. And the rabbits arrived with few parasites or diseases. There were some native predators when the rabbits arrived, including dingos, native cats (*Dasyurus* species), and wedge-tailed eagles, but extensive predator control in Victoria and New South Wales around the same time that rabbits began their long march reduced their potential impact. Experts believe, however, that the single greatest ally of the rabbits was changes in land-cover vegetation that accompanied the introduction of domestic livestock. To better raise sheep and cattle, ranchers replaced the natural trees, shrubs, and perennial tall grasses with introduced annual herbs and grasses preferred by livestock. This created perfect grazing land for rabbits, too.

Rabbits have contributed to the decline of several native Australian marsupials, including the greater bilby. The rabbits compete with bilbies for burrows, and bilbies inadvertently are killed in traps and by poison bait set out for rabbit control. Ironically, the bilby, whose very long ears resemble rabbit ears—it species name, *lagotis*, means "hare ears," has been touted in Australia as a replacement for the Easter Bunny, with chocolate versions sold to raise funds for the bilby's conservation.

Australia struggled mightily with its rabbit problem until the myxoma virus was deliberately introduced in 1950. In the initial epidemic, 99% of Australia's rabbits died, and they disappeared altogether in some marginal habitats. But the virulence of the myxoma virus declined quickly and rabbits recovered, although never to previous numbers. Estimates suggest that rabbit numbers in the early 1990s were between 5 and 25% of what they were before 1950, depending on the habitat, with the larger number sur-

A section of one of Australia's rabbit-proof fences, photographed in the 1920s, which stretched for thousands of kilometers in an attempt to keep rabbits out of pasturelands.

Photo from Library and Information Service of Western Australia

viving in arid rangelands and fewer in high rainfall areas (see "Do rabbits get sick?" in chapter 5).

Rabbit numbers fell dramatically again after the calicivirus escaped in 1995 from a scientific laboratory on an island off South Australia and was subsequently deliberately reintroduced. The calicivirus causes rabbit hemorrhagic disease (RHD), which kills most infected rabbits within days of exposure. Those that survive a bout of RHD are immune, however, and females pass that immunity to their young, which are then resistant to the disease until they are about 13 weeks old. So, rabbits remain and even in their currently reduced numbers are responsible for economic losses amounting to millions of dollars a year annually, combining the costs of rabbit control and loss of agricultural productivity, without even counting the environmental impacts, which are substantial.

In contrast to this, domestic rabbits are coming to the rescue of the endangered Scott's tree kangaroo (*Dendrolagus scottae*), also called the tenkile, in Papua New Guinea. The tree kangaroos were decimated in recent years by villagers who hunted the slow-moving marsupials for food. Conservationists encouraged villagers to breed domestic rabbits as a substitute, and it's making a real difference: the tree kangaroos have been sighted in places they hadn't been seen in years. The particular breed of rabbit is rather sluggish, posing little escape risk, and should there be escapees, they are very unlikely to survive on their own in Papua New Guinea's hot, humid rain forest.

Setting aside the immense problems caused the historically recent introductions of European rabbits to far-flung corners of the globe, other lagomorphs are sometimes considered pests although we should bear in mind that no animals are inherently pests—human activities and attitudes make them so.

Take the case of the Afghan pika in Baluchistan, Pakistan. Until recently, only nomadic herders used the high mountain valleys of this region that the pikas inhabit. But population growth led to some people taking up farming of subsistence crops such as barley and wheat. In addition, apple trees grow well here so apple orchards were developed as a cash crop. This cornucopia of food proved very attractive to pikas, especially in the winter when other green plants are scarce. While relatively minor pests of grain and vegetables, they do serious damage in apple orchards, resulting in serious economic losses. As a result, various control programs, including poisoning, are in place there. But people contributed to the problem. They built rock walls to surround the orchards and left rock piles nearby after clearing the fields—both ideal habitat for Afghan pikas. What's more, large herds of free-ranging domestic sheep and goats scour the landscape of its green vegetation throughout the warmer months, leaving the pikas hard-pressed to build up the hay piles they need to survive the winter. Elsewhere, it has been reported that livestock are attracted to pika hay piles, so it is possible that even as pikas build a pile it is being eaten away!

Other species of pikas have long been considered pests in China, where it is believed they contribute to degradation of livestock grazing lands. Pikas have been poisoned on a massive scale there, despite there being no evidence they truly have any detrimental effects. Indeed, like prairie dogs in the western United States, which have been similarly subjected to extensive control operations, evidence suggests quite the opposite (see "Are rabbits good for the environment?" in chapter 5). Pika control in some parts of Tibet has even contributed to an increase in brown bear conflict with people. These bears, for which pikas are a dietary mainstay, are increasingly entering human habitats to kill livestock.

In the western United States, black-tailed jackrabbits are considered serious pests, both because they compete with livestock for food and because they eat a wide variety of crops and, like Afghan pikas, damage orchard trees. According to various estimates, 148 black-tailed jackrabbits eat as much forage as one cow and just six jackrabbits as much as a sheep. Among the crops they eat or damage are vegetables including carrots, beans, and lettuce; fruits and berries such as apples, cherries, and raspberries; herbs; and ornamentals including turf grass and flowers. Black-tailed jackrabbits are adaptable and live easily among people; what's more, some human activities favor jackrabbits. Overgrazing by cattle creates good habitat for black-tails, which prefer open weedy areas to tall grass. Crops, obviously, are attractive foodstuffs and especially during droughts, irrigated croplands are magnets for hungry jackrabbits. Black-tailed jackrabbit numbers also fluctuate dramatically in approximately 5- to 10-year cycles. At peak numbers, jackrabbit densities of nearly 500 and up to 1,500 per square kilometer (1,200 to 3,600 per square mile) have been reported. When a single jackrabbit can eat 1.1 to 2.2 kilograms (0.5 to 1 pound) of green vegetation each day, the potential for damage to crops is clearly enormous.

The early settlers trying to farm in the western United States had to contend with large-scale black-tailed jackrabbit depredations, as their very act of farming spurred massive growth in jackrabbit numbers, perhaps exacerbated by people also controlling predators. In some areas, jackrabbits reportedly ate all the crops! In the late 1800s, farmers attempted to prevent crop losses with huge rabbit drives. Hundreds or even thousands of farmers in a district joined in these drives, with people surrounding an area several miles wide and then driving the rabbits toward fences that funneled into a large corral. Once corralled, the rabbits were clubbed to death. The largest of these drives, one near Fresno, California, in 1892, involved about 6,000 people and the death of about 20,000 rabbits. Such drives were largely abandoned by the turn of twentieth century, as farmers turned to other means of control such as hunting, fencing, use of chemical repellents, and poisoning. Still, farmers in Kansas revived the practice in the 1930s and again in the 1950s, when drought years were correlated with an increase in black-tailed jackrabbits to "near plague proportions."

Jackrabbits have also proven to be problematic at some airports, whose expanses of open vegetated spaces attract them. While jackrabbits may be run over by airplanes on runways, they themselves pose little threat to the aircraft. Dogs, however, chase them and collisions with dogs may damage planes. More important, however, are the predatory birds hunting jackrabbits. Birds may be sucked into jet engines and collide with windows. Thus, jackrabbits are often controlled at airports in the western United States.

There was an interesting situation in the Miami International Airport in 2003. About 500 black-tailed jackrabbits had taken up residence there, although how they got there is not known. Jackrabbits are not native to Florida, and authorities were concerned about their escaping the confines of the airport and invading other Florida habitats. In addition, jackrabbits were sometimes killed in collisions with planes or other vehicles, attracting turkey and black vultures (*Cathartes aura* and *Coragyps atratu*) to scavenge on the carcasses. In 2 years, from 2001 to 2003, 24 vultures were hit by planes, posing a safety hazard as well as requiring costly repairs to planes. Compelled in part by the Federal Aviation Administration, this jackrabbit population was eradicated by a combination of lethal control and live-trapping and translocation to Texas.

A rabbit drive in Hermiston, Oregon, some time between 1910 and 1915, killed large numbers of crop-eating jackrabbits. Photo from Library of Congress, Prints & Photographs Division, LC-DIG-ggbain-12494

Black-tailed jackrabbits may attract vultures to airports, where bird strikes are a serious problem for planes.

Photo by Scott Rheam, USFWS

How can I keep rabbits out of my garden?

Don't go into Mr. McGregor's garden: your Father had an accident there;
he was put in a pie by Mrs. McGregor.

Beatrix Potter, The Tale of Peter Rabbit

Most experts agree that the best way to keep rabbits out of a garden,
where they will eat just about anything green that you try to grow—is with
a fence that excludes them. The fence should be fine mesh, such as quarter-
inch hardware cloth, 2 to 4 feet high, and buried several inches below the
soil surface. Leave a foot or two between the fence and the plants.

It also helps to remove any habitat you may have inadvertently created
for rabbits, such as wood or stone piles and patches of tall grass or weeds.
All of these offer rabbits cover and nesting sites and without them rabbits
will move away.

There are a variety of commercially available chemical repellents, including those that smell like carnivore urine or feces, but their effective-
ness varies. They also suffer from the fact that they don't last very long so
you must apply them repeatedly as your plants grow and at least after every
rainfall or watering. The same applies to the use of "natural" deterrents,
such as garlic juice, black pepper, or red pepper spray, or scattering your
cat's used litter material around the edge of the garden.

A pet dog or cat that patrols your garden is also useful in deterring rabbits.

Are rabbits bad for lawns?

Rabbits and hares do eat lawn grasses but only sometimes in sufficient
numbers or quantity to harm your lawn. If you are concerned about rabbit damage to your lawn, you can use the methods noted above to eliminate or reduce their impact. That said, lawns are not very environmentally friendly. In the United States, significant amounts of water are used
to maintain lawns. Gas- or diesel-powered power mowers are expensive
and contribute to carbon emissions and other forms of pollution including noise pollution. Herbicides, insecticides, and fertilizers (mostly based
on fossil fuels) used to keep lawns green and weed free are pollutants, and
run-off contaminates land and water far from your own property. If wildlife
is to survive in our increasingly human-dominated landscapes, people will
have to develop more tolerant attitudes and leave places for animals to live.
Instead of worrying about rabbits damaging your lawn, think about the way
you care for your lawn might be harming rabbits. If you live where rabbits
are abundant, consider landscaping with hardy small green plants that rab-

bits like to eat and let the rabbits act as natural lawn mowers and fertilizers. You'll save money, help protect the environment, and have the pleasure of watching these fascinating animals from your window.

Are rabbits dangerous?

Readers of a certain age may remember the 1979 "killer rabbit" incident involving then-president Jimmy Carter. While fishing on a pond on his farm in Georgia, Carter had to fend off a swamp cottontail apparently determined to board his small boat. A White House photographer snapped a picture of the president shooing the animal away with an oar. What was going on with this rabbit is unclear—it was described as hissing and gnashing its teeth—but this may be the only documented case of a rabbit or hare attacking a person without provocation. And even then, the provocation must be up close and personal, such as actually handling a live rabbit, which may try to kick, bite, or scratch its handler.

Wild rabbits are associated with transmitting a handful of diseases to people. The most important of these is tularemia, also known as rabbit fever. Tularemia is caused by the bacterium *Francisella tularensis*, which is one of most highly infectious bacteria known—it takes only from 10 to 50 of these organisms to cause an infection in people. Rabbits and hares in North America and hares in Eurasia (along with many rodents species) are susceptible to and carry tularemia, which may be spread either by a vector, usually a tick, from lagomorphs to people, or via direct contact between an infected lagomorph and a person. *F. tularenis* can also spread via aerosolization and may be inhaled directly into the lungs, where it causes the most dangerous form of the disease, a pneumonia that may be fatal in untreated patients.

Most likely to be infected are people who handle lagomorphs, such as hunters, or who ingest meat from an infected animal. But certain activities in the vicinity of an infected animal may also increase risk. For example, there have been cases of people getting the disease after cutting brush or mowing lawns where infected animals live, because these activities mechanically aerosolize the bacteria.

Tularemia is endemic in parts of North America and Eurasia. There are intermittent localized outbreaks, but the incidence of the disease has dramatically declined in the United States and Russia, where the best data exist, since the middle of the last century. Mild cases of the disease may be underreported, though. It is not tested for in standard lab blood tests and although it can usually be treated successfully with antibiotics, it may not be diagnosed. Tularemia does not spread from person to person, but house cats can become infected if they eat dead rabbits or infected rodents and transfer the infection to their human caretakers.

A swamp rabbit, seen swimming in the right side of the photo, appeared to be threatening then-president Jimmy Carter, who shooed it away with a splash of water. Photo courtesy of the Jimmy Carter Library and Museum

Is it safe to eat rabbits?

Hunting cottontails and jackrabbits for food is very popular in parts of the United States, as is hunting rabbits and hares for food around the world. In general, it is safe to eat wild rabbits or hares but an animal that acts strangely before it is shot or captured should not be eaten. White spots on the liver indicate that an animal may have tularemia, and animals with this symptom should be discarded, too. The bacteria that cause tularemia are killed by cooking the meat to well-done but not by freezing.

The meat of domestic rabbits that you can purchase, usually frozen, in supermarkets in the United States is as safe as the chicken and other meats sold here.

What should I do if I get bitten by a rabbit?

It's hard to imagine the circumstances in which the average person would get bitten by a wild lagomorph! Lagomorphs treat approaching humans like they do any other potential predator and run away, but a rabbit or hare that does not try to escape is likely to have some sort of illness. In any case, being bitten by any wild animal is potentially dangerous and you should seek medical care immediately. Lagomorphs are rarely diagnosed with rabies, although they can become infected with this disease, probably because most encounters with a rabid carnivore close enough for a bite to be inflicted are fatal to lagomorphs. Still, specialists recommend that testing for rabies following a bite from a lagomorph should be considered on a case-by-case basis.

Chapter 10

Human Problems (from a rabbit's viewpoint)

Are any rabbits endangered?

Many people are surprised to learn that some rabbits are endangered. How could animals proverbial for profligate reproduction and considered serious pests in some places be endangered? As we've shown elsewhere, not all lagomorphs breed like rabbits, and even the European rabbit, on which this expression is based, is now in trouble. But, more important, even profligate reproductive capabilities don't necessarily compensate for the mostly human-caused environmental changes that lead to species endangerment. In fact, low rates of reproduction, such as those famously associated with giant pandas, are not usually a cause of endangerment, although they make it harder for a species to recover from declines due to other factors.

Taken all together, the reasons many lagomorphs are declining are similar to the reasons so many other animals are declining: loss, degradation, and fragmentation of habitat; conversion from traditional to industrial agriculture; excessive killing and consumption by people; invasive species (including other species of lagomorphs); disease; and climate change.

The International Union for the Conservation of Nature (IUCN) Red List of Threatened Species is the authoritative source of information on the status of species. Its assessments use rigorous scientific criteria and are based on the input of many experts who serve on the specialist groups of the IUCN's Species Survival Commission. The most recent assessment of lagomorphs was completed in 2008, under the leadership of Andrew Smith, chair of the Lagomorph Specialist Group.

Two lagomorphs, the riverine rabbit and the silver pika, are considered "critically endangered," 10 are "endangered," and five are "vulnerable."

Together, animals in these three categories are considered "threatened" with extinction. Another six species are listed as "near threatened," and eight fall into the unhappy category of "data deficient"—which means that specialists have no idea how these species are faring. Adding these numbers to the threatened category—not an unreasonable thing to do given the gloomy downward trajectories most mammals are facing—means that fully one-third of the named lagomorphs are in some kind of trouble. And, if additional species are named (see "How many species or rabbits are there?" in chapter 1), the situation will get worse not better. For instance, the mountain hare is listed as of "least concern," but if the Irish hare subspecies is elevated to full species status, its conservation status may be downgraded because its numbers are small and rapidly declining. Similarly, as new cottontail species are described in Central and South America, they may be found to be threatened given the extent of land-use change and habitat destruction in this region.

Looking at the population trends of the rest, the 62 species deemed of "least concern," doesn't make the picture any rosier. Only the eastern cottontail is considered to be increasing; 12 species are called stable and another 12 are decreasing. For a whopping 37 species—more than a third of known lagomorphs—no information is available to tell whether their numbers increasing or decreasing. This list includes many pikas living in remote corners of China as well as mountain cottontails and marsh rabbits in the United States.

The Red List assessments do not address subspecies, unlike the U.S. Endangered Species List, so whatever the overall status of the marsh rabbit, for instance, we know that the Lower Keys subspecies (*Sylvilagus palustris hefneri*) of south Florida is considered critically endangered. The Columbia Basin pygmy rabbit population, which was found in Washington state, is almost certainly extinct in the wild although the species is assessed as of least concern but with declining numbers. The U.S. list also includes the riparian brush rabbit (*S. bachmani riparius*), a subspecies found only in California's San Joaquin Valley. Also in the United States, the status of the New England cottontail, the pygmy rabbit, and the white-sided jackrabbit is under review for listing by the Fish and Wildlife Service under the Endangered Species Act (ESA). A petition to list the American pika was denied in early 2010.

Many species are protected on other nations' equivalents of the ESA, but most often this is paper protection with few or no on-the-ground conservation programs or effective enforcement of antipoaching laws and the like. Even where threatened species live in national parks and protected areas, enforcement of laws against human encroachment and poaching are often lax. Especially in the developing world, national budgets rarely are sufficient

Endangered and threatened lagomorphs species

Species	IUCN Red List status	U.S. Fish and Wildlife Service status
pygmy rabbit (*Brachylagus idahoensis*)		Under review
Columbia Basin pygmy rabbit (*Brachylagus idahoensis*)		Endangered
riverine rabbit (*Bunolagus monticularis*)	Critically endangered	
bristly rabbit (*Caprolagus hispidus*)	Endangered	
white-sided jackrabbit (*Lepus callotis*)		Under review
broom hare (*Lepus castroviejoi*)	Vulnerable	
Corsican hare (*Lepus corsicanus*)	Vulnerable	
Tehuantepec jackrabbit (*Lepus flavigularis*)	Endangered	
Hainan hare (*Lepus hainanus*)	Vulnerable	
Sumatran striped rabbit (*Nesolagus netscheri*)	Endangered	
silver pika (*Ochotona argentata*)	Critically endangered	
Hoffmann's pika (*Ochotona hoffmanni*)	Endangered	
Ili pika (*Ochotona iliensis*)	Endangered	
Kozlov's pika (*Ochotona koslowi*)	Endangered	
Amami rabbit (*Pentalagus furnessi*)	Endangered	
volcano rabbit (*Romerolagus diazi*)	Endangered	
riparian brush rabbit (*Sylvilagus bachmani riparius*)		Endangered
Tres Marías cottontail (*Sylvilagus graysoni*)	Endangered	
Omilteme cottontail (*Sylvilagus insonus*)	Endangered	
Lower Keys marsh rabbit (*Sylvilagus palustris hefneri*)		Endangered
robust cottontail (*Sylvilagus robustus*)	Endangered	
New England cottontail (*Sylvilagus transitionalis*)	Vulnerable	Threatened

for effective conservation of wildlife. Moreover, threatened lagomorphs simply haven't attracted the international attention that is devoted to more high-profile, charismatic megavertebrates such as tigers and elephants.

Strong, active efforts to conserve and restore European rabbits in Spain and Portugal are under way, and significant research is being devoted to understanding the decline of European hares in Western Europe and Irish hare in Ireland. The Amami rabbit has been declared a Japanese National Monument. It is protected from hunting and also lives in some national protected areas. Attempts have also been made to reduce numbers of the invasive predatory mongooses that pose a major threat to this species. In South Africa, a large number of organizations have united in an attempt to save the riverine rabbit, which numbers fewer than 250 adults in 10 iso-

Rabbits: The Animal Answer Guide

The Lower Keys marsh rabbit, which has been isolated from its mainland counterparts for about 10,000 years, is critically endangered.

Photo © Jason Schmidt

lated populations; this effort is coordinated by the Endangered Wildlife Trust's Riverine Rabbit Working Group.

When a species is listed in the United States, the law requires the development and implementation of a recovery plan, and these exist for the riparian brush rabbit, the Lower Keys marsh rabbit, and Columbia Basin pygmy rabbit. For instance, when the riparian brush rabbit was listed in 2000, the subspecies was down to a handful of individuals in the only two fragments of habitat remaining for them in the San Joaquin Valley's agriculture-dominated landscape. One of the fragments is the 1.02 square kilometer (253-acre) Caswell Memorial State Park; the other is about 1.07 square kilometers (270 acres) of private land. Since then, under the leadership of Patrick A. Kelly of California State University, Fresno, the rabbit has been plucked from the brink of extinction through a comprehensive conservation effort that has included research and monitoring, captive breeding, and release back into the wild to establish a new population in the San Joaquin National Wildlife Refuge and on private land.

The New England cottontail is threatened, in part, by competition from bolder eastern cottontails.

Photo by Anne Terniko, USFWS

Conservation of threatened species often begins with one or a few champions of a particular species. John Litvaitis of the University of New Hampshire has long been a champion of the New England cottontail. In part due to his efforts, the species can no longer be hunted in New Hampshire and Maine, and in both states there are active recovery programs.

Will climate change affect rabbits?

There has been much debate about the causes and potential impacts of global warming, but scientists around the world agree that global climate change is real and the result of human activities. This is the consensus view of hundreds of scientists and the governments of about 100 countries, including that of the United States, according to the Intergovernmental Panel on Climate Change. Average global temperatures have increased since the mid-twentieth century, very likely due to increases in greenhouse gas concentrations that arise from human activities. This is called the "greenhouse effect." Warming of the oceans, rising sea levels, glaciers melting, sea ice retreating in the Arctic, and diminished snow cover in the Northern Hemisphere are evidence of global warming. Some changes in plants and animals believed to be effects of climate change are already evident as well.

The most important greenhouse gas directly affected by human activities is carbon dioxide. The amount of carbon dioxide in the atmosphere has increased by about 35% in the industrial era, primarily due to burning fossil fuels and cutting down forests. Methane, nitrous oxide, ozone, and several other gases present in the atmosphere in small amounts also contribute to the greenhouse effect. Water vapor is the most important greenhouse gas, but human activities have little direct affect on changes in water vapor. However, as oceans warm (a direct affect of human activities), water vapor in the atmosphere increases. This increased water vapor, in turn, is changing patterns of precipitation (rainfall and snowfall). In some regions, the change means wetter conditions and more floods; in other parts of the world, it has the opposite effect: drier conditions and droughts and increased wildfires. More precipitation now falls as rain rather than snow in northern regions.

Coral reefs, northern forests, mountains, coasts, Mediterranean-climate areas, and polar sea ice are likely to be most severely affected, but all areas face significant changes. Many natural systems are being affected by climate change, particularly temperature increases, but these affects vary regionally.

How much global average temperatures will increase by the end of this century is uncertain and will depend, among many other things, on the extent to which we find and use ways to reduce emissions of greenhouse gases. However, scientists believe the increase is likely to be in the range of 2 to 4.5°C (3.6 to 8.1°F) with a best estimate of about 3°C (5.4°F), and is very unlikely to be less than 1.5°C (2.7°F). Broadly, 20 to 30% of plant and animal species assessed are likely to be at increased risk of extinction if increases in global average temperature are more than 1.5 to 2.5°C (2.7 to 4.5°F). This is a rough estimate however and may vary regionally from 1 to 80%. If the temperature increases are greater than 2.5°C, scientists predict major changes in ecosystems and in species' ranges and interactions with other species, with mostly negative effects on biodiversity and likely significant extinctions around the world.

Other contributors to species endangerment—such as habitat fragmentation and loss, invasive species, disease, and pollution—can increase a species' vulnerability to climate change. For instance, a population living in an isolated reserve surrounded by impassible human development may not be able to move to another area to escape one that climate change has made unsuitable. Moving, however, is a viable adaptation to climate change in some species, although this may have a negative effect on the species already living in the new area. Land-living species are moving toward the cooler poles or moving to higher, cooler elevations. In the Northern Hemisphere, the timing of natural events such as flower blooming, insect emergence, and bird migration has changed. Many migratory bird species are arriving on the breeding grounds either earlier or later in the spring. Some small mammals are coming out of hibernation and breeding earlier.

So, what does this all mean for the world's lagomorphs?

Most immediately at risk are the pikas, which have all of the characteristics that scientists predict make some species more vulnerable than others. They have small natural ranges in mountains at high elevation, referred to as mountain or sky islands (see "Which rabbits have the largest distributions and which the most restricted?" in chapter 5). Their special-ized habitats are patchily distributed among hostile terrain; they tend not to move very far; and physiologically they cannot cope with high tem-peratures. American pikas, for instance, die within a few hours if they are exposed to temperatures of about 35.5°C (78°F) or higher. There is very good evidence that American pika populations are already going extinct at lower elevations. In fact, the American pika was the first species proposed for listing on the U.S. Endangered Species List whose endangerment is at-tributed to global warming, although the proposal was denied.

One way scientists attempt to predict the future impacts of climate change is to look at how climate change in the past affected a species or ecosystem. This has been done by Donald Grayson for American pikas living in the Great Basin of western United States. From about 40,000 to 7,500 years ago, the average elevation at which pikas lived in the Great Ba-sin was 1,750 meters (5,741 feet). From 7,500 to 4,500 years ago, however, the climate became hotter and dryer with attendant vegetation changes, and the average elevation of pika populations rose to 2,168 meters (7,113 feet), and lower-elevation populations went extinct. Today, according to research by Eric Beever, the average minimum elevation of the 18 Great Basin populations is 2,533 meters (8,310 feet), with a 100-meter (328-foot) increase documented in the past decade alone. Pikas have disappeared from nine of 25 locations where they were found historically. The effects of in-creasing temperatures in the Great Basin are also exacerbated by increased human impacts from roads and related habitat degradation.

There is also evidence that collared pikas, native to western Canada and Alaska, are suffering from climate change. David Hik found that collared pikas declined by 90% during the winters of 1999 and 2000, when warm-ing caused snowmelt, rain, and refreezing in the middle of winter. This resulted in the pikas losing the insulation normally provided by snow and the vegetation they need for food being buried under ice. With glaciers retreating in the Himalayas under the influence of higher temperatures, it is very likely that Asia's pikas, most of which live on the Tibetan Plateau will, at best, decline as they must move higher upslope. And the Ili pika of northwest China, a species that was just described in 1986, may have already succumbed to warming, perhaps compounded by the fact that hu-man herders are also grazing their livestock at higher elevations as warmer temperatures allow vegetation to grow at greater altitudes.

Historical data also show that climate change is likely to further con-strict the range of pygmy rabbits. In the Great Basin, for instance, at the same time that America pikas declined due to warmer and drier climate between 7,500 and 4,500 years ago, so did pygmy rabbits, as this climate favored pine-juniper woodlands over the big sagebrush habitat they spe-

American pikas are moving to higher elevations as temperatures warm in the Great Basin of western United States. Photo by Walter Siegmund (with permission), Wikimedia Commons / CC-BY-SA 3.0

cialize on. In the last century, the elevational range of pygmy rabbits has shifted 220 meters (722 feet) upward. There are many reasons for the massive decline in sagebrush habitat, including not only the encroachment of pine-juniper woodlands but also massive replacement of the sagebrush with farm and grazing lands and other human development activities. All of this has led to the shrinking range of pygmy rabbits. But warming has other effects, too. These small rabbits used tunnels under the snow to dine on sagebrush in winter; in the tunnels, they are also hidden from predators. Greater predation pressure may account for the rabbits disappearing from warmer, lower-elevation sites with less snow cover.

The volcano rabbit, a species confined to pine forests and alpine grasslands on a few high-elevation mountain slopes in Mexico, is another potential victim of climate change. Temperatures in their habitat have increased over the past 50 years as has precipitation during once-dry winters. Ecological predictions suggest that there will be significant reductions in the volcano rabbit's distribution at lower elevations, again compounded by more intense farming and livestock grazing.

Warming is also affecting the distribution of mountain hares both in glacial refugia (isolated patches) in the Alps and in their northern Scandinavian range. With warming, European hares are expanding into these regions. This threatens mountain hares, because European hares outcompete

Human Problems

Pygmy rabbits dig their burrows in specific soil types in sagebrush habitat, which has almost disappeared. Photo © Jim Witham

them for habitat. In addition, these two species may interbreed when their ranges meet, so there is a risk of genetic "pollution." A similar situation may be occurring in the northeastern United States, where eastern cottontails are expanding northward to the detriment of New England cottontails. In both of these cases, the expansion of the more southerly species has been abetted by human translocations.

For very different reasons, climate change is one of many serious threats to the survival of Florida's Lower Keys marsh rabbit. This species' decline is largely due to habitat loss, and the rising sea levels that are predicted as a result of climate change threaten to inundate much of the habitat that remains to them. Unless the rate of sea level rise is matched by the rate at which vegetation can migrate inland to create compensatory habitat, the future looks bleak for this critically endangered subspecies. This will be determined by whether human activities that prevent the inland movement of habitat are curtailed or abandoned. Although less critical in the near term, habitat for marsh rabbits throughout their coastal range on the southeastern Atlantic and Gulf coasts may be affected by rising sea levels in the decades ahead.

Scientists are also concerned about the potential mismatch between the timing of the snowshoe hare's coat color changes and that of the onset and end of winter snow cover. Coat color changes are dictated by changing day length: more hours of darkness trigger the change from brown to white fur in the fall and the reverse happens in the spring (see "Do fur colors change in different seasons?" in chapter 3). But snow cover depends primarily on temperature, so later snow cover in the fall and earlier loss of snow cover

Rabbits: The Animal Answer Guide

in the spring may leave the hares "thinking" their white pelage is hiding them from predators when in fact they are particularly conspicuous against ground devoid of snow. This may also present a problem for the other hares whose fur turns white during the winter.

There is also evidence to suggest that warming may influence snowshoe hare's ability to escape from predatory Canada lynx. When there are few warm spells during the winter, the snow stays fluffy. Lynx sink into fluffy snow, hampering their ability to chase prey, while hares are able to run on the snow surface and thus escape their pursuers more easily. More warm spells, however, mean the snow surface forms a crust after melting and re-freezing. This would give lynx an advantage because they would be able to run on this hard surface without sinking. This change could disrupt the finely tuned hare-lynx cycle of the North American boreal forest (see "Are rabbits good for the environment?" in chapter 5).

Are rabbits affected by pollution?

Apart from the effects of carbon emissions on global climate change, there is little information on the impact of pollution on wild lagomorphs. Much research has been conducted to test the effects of chemicals, such as pesticides, and other potential pollutants on laboratory rabbits. In fact, the U.S. Environmental Protection Agency mandates that most pesticides be tested for toxicity and other effects in rabbits, but this is to determine their effects on human health rather than on wildlife. It is hard to extrapolate from these laboratory tests to the wild situation, but it seems safe to say that chemical and other pollutants have some effect on lagomorphs. Some studies have tried to determine whether lagomorphs could act as biological indicators of certain pollutants in the environment, again with an eye to human health and safety. For instance, hares in Eastern Europe were shown to be sensitive bioindicators of mercury pollution in the environment.

In 1943, the Handford Site in the Columbia Basin of eastern Washington was set aside to produce plutonium for the Manhattan Project of the U.S. Department of Defense; plutonium from this site, which encompassed an area of 1,450 square kilometers (560 square miles), was used to make the atomic bomb dropped on Nagasaki in Japan in 1945, ending the Second World War. At its peak in 1964, nine nuclear reactors were in operation at Hanford but all were shut down by 1989. While the reactors were running, large amounts of radioactive waste were released into the air, ground, and water, and nuclear-waste storage facilities continue to leak waste into the environment despite billions of dollars spent on cleaning up the site. Black-tailed jackrabbits and mountain cottontails are among the several mammals that have been monitored to study radioactive contamination on the site.

In various studies, jackrabbits and cottontails have been found to accumulate low levels of plutonium and other radioactive materials. At least one study showed that the behavior of irradiated jackrabbits differed from that of nonirradiated jackrabbits. However, there are no data on reproduction or longevity or related parameters to determine any potential long-term effects of radioactive contamination. The Columbia Basin pygmy rabbit had also lived on the site until they went extinct in 1985 following a brushfire that consumed their small patch of habitat. Overall, however, the lagomorphs there appear to thriving. Because the site has been off-limits to all other human activities and 90% of it was not used for reactor operations, it contains one of the largest remaining pieces of steppe sagebrush habitat in the country. More than half of the site's land was declared the Hanford Reach National Monument / Saddle Mountain National Wildlife Refuge in 2000.

One surprising fact is that wildlife has appeared to thrive at other sites contaminated by nuclear material, which scientists attribute to absence of other human impacts. For instance, surveys around Chernobyl, in Ukraine, site of the worst nuclear disaster ever, showed more abundant hares, deer, wild pigs (*Sus scrofa*), and other wildlife within the highly radioactive exclusion zone set up after the 1986 event than are found outside that zone. This is despite the fact that radioactive materials deposited on vegetation build up in the tissues of herbivores that ingest the contaminated plants. A study conducted in Finland more than 15 years after Chernobyl showed that mountain hares still had high concentrations of radiocesium from the disaster's far-reaching fallout. However, long-term, detailed studies have not been conducted to determine whether exposure to radioactive materials has affected the genetics, reproduction, or other features of hares or other mammals, although there is some research on birds that suggests wildlife may be suffering in subtle ways. Barn swallows (*Hirundo rustica*), for instance, have poorer body condition, reproduce less frequently, and produce smaller clutches of eggs and smaller broods of hatchlings in highly contaminated areas of Chernobyl.

Glyphosate is a widely used herbicide, best known as the active ingredient in Roundup. In an experimental study of snowshoe hares in British Columbia, the application of glyphosate to forest habitat appeared to have no effect on the hares' reproduction or longevity. Snowshoe hares have also been found to avoid equally areas "clear-cut" by herbicides and those clearcut manually. In another study, mountain hares avoided areas treated with herbicides but only for a year. However, chronic exposure to glyphosate has been shown to decrease body weight in laboratory rabbits. One study on male laboratory rabbits found high dosages of the chemical to have multiple affects on reproduction, including decreases in sperm concentration and increases in abnormal sperm.

In analyzing the threats to Lower Keys marsh rabbits, the U.S. Fish and Wildlife Service determined that chemical poisons used to kill rats posed a serious risk to the species' survival because of the potential of the rabbits ingesting the toxic bait. There is also concern about this water-loving rabbit's potential exposure to pesticides in marsh habitat and degraded water quality due to run-off from septic tanks and fertilizer.

Herbicides may also play a role in the decline of the European hare in Europe. Hares have been poisoned by acute ingestion of herbicides, but there is no evidence that this did or did not result in any long-term effects on the health or reproduction of these animals. Experts believe, however, that it is quite likely that the extensive use of herbicides in agricultural fields did have an indirect impact on the hares' numbers. Herbicides were first used in the late 1950s and their application has increased since then; this time frame coincides with that of the hare's decline. What is perceived to have helped the human farmer or gardener may have hurt the hares (see "Are rabbits bad for lawns?" in chapter 9). Scientists found that herbicides have reduced the abundance of weeds in cereal crop fields by a factor of at least 10, thus significantly reducing their availability to hares, whose summer diet largely consists of a variety of weeds. The use of herbicides and pesticides has also been mentioned as contributing to the decline of European rabbits in Spain and Portugal.

Why do people hunt and eat rabbits?

Rabbits and hares make a good meal; it's as simple as that. Archeologists have evidence of people hunting rabbits in the south of France 120,000 years ago; undoubtedly, hominids were eating lagomorphs earlier than that. A recent study found that during the last ice age, about 40,000 to 10,000 years ago, when the Iberian Peninsula was a refugium (an isolated area that forms a natural refuge) from the glaciers that covered much Europe, early modern humans and the last of the Neanderthals may have survived on diets largely made up of rabbits. There is abundant archeological evidence from North America, southern Europe, and northern Africa that our ancestors turned to rabbits and hares for subsistence whenever preferred large game, such as deer, or slower more easily captured prey, such as turtles and shellfish, became scarce due to overhunting or climate change. While lagomorphs may be small, they are abundant, and their rapid reproduction means that their populations recover quickly from hunting, providing a more reliable source of food. Lagomorphs may also have been found closer to settlements, reducing the travel time associating with hunting large game. With the advent of farming, crops attracted lagomorphs to settlements too.

Research on animal remains from Paleo-Indian sites throughout the western United States from roughly 12,000 years ago shows that cottontails and jackrabbits formed a small but significant portion of a diet otherwise dominated by mammoths, bison, and other large mammals. And since that time, lagomorphs played a more or less important role in human subsistence, their numbers in archeological remains generally being in inverse proportion to those of artiodactyls such as deer and sheep. Remains identified as eastern cottontails have been found in sites in Belize that date between 1000 and 400 BCE and in Maya settlements that date to 200 CE and to 1500 CE in Mexico. The Anasazi—the ancient Pueblo people of what is now the southwestern United States—were large consumers of cottontails and jackrabbits until their culture began to disappear in the twelfth century CE. Both cottontails and jackrabbits were captured in large communal drives, and cottontails, which took advantage of the food provided in agricultural fields, were taken opportunistically as well as in what archeologists call "garden hunts." In the eastern United States, Native Americans used fire to clear patches in forests that attracted deer, cottontails, and other food animals for harvesting.

Snowshoe hares remained important to the subsistence of Native Americans living in boreal forest into the twentieth century. In a detailed ethnographic account of Chipewyan food and hunting habitats in the boreal forest in Canada, scientists found that while moose formed the largest part of the diet by weight, snowshoe hares were trapped and eaten on daily basis. Men hunted moose and other animals, but women were the hare trappers, setting snares and checking their trap lines every day as part of their overall responsibility for provisioning and preparing food for their families. In the winter, hare carcasses were left to freeze and thawed out as needed for dinner; in summer, hares were butchered for immediate consumption or the meat was preserved by smoking.

Almost wherever rabbits and hares occur, they appear to form some part of the human diet. For example, the Bunyoro rabbit is hunted as bushmeat in the Congo, the Hainan hare is hunted for meat on that Chinese island, and the Annamite striped rabbit was discovered in a meat market in Laos. There is no evidence of people eating pikas, but it would be surprising if they were never eaten at all. Eating rabbits and hares, however, is taboo in some cultures. Jewish dietary laws dictate that rabbits and hare are not kosher; Leviticus 1:11 states, "And the hare, because he cheweth the cud, but divideth not the hoof; he is unclean unto you. And thus cannot be eaten." Eating rabbit or hare is permitted in Sunni but not in Shia Islam.

The meat of rabbits and hares is high in protein and low in fat and cholesterol. This may seem like a healthy food choice today but can be a drawback is some situations. Trying to survive on a diet of such lean meat without adding another source of fats and carbohydrates can actually lead to what is known as "rabbit starvation," a form of malnutrition that causes, among other symptoms, diarrhea, lassitude, and headaches that can only be cured by consuming fat or carbohydrates. The human body cannot obtain enough energy in the form of glucose by breaking down protein alone, especially under conditions of cold stress; carbohydrates and fats are required. A lean meat diet is also deficient in some essential fatty acids.

Explorers Lewis and Clark reported symptoms of rabbit starvation in their diaries when their diet was restricted to game. Charles Darwin, in *The Voyage of the Beagle*, mentions the syndrome (without using the name) in reporting that the gauchos of South America could live on an all-meat diet because they ate a prodigious amount of fat.

The term itself was popularized by Arctic anthropologist and explorer Vilhjalmur Stefansson, who wrote the following in his 1938 book *Unsolved Mysteries of the Arctic*:

> A well known field where such deaths occur is northern edge of the forest in Canada where Indians are sometimes unable to find any food except rabbits. The expression "rabbit starvation," frequently heard among the Athapasca Indians northwest of Great Bear Lake, means not that people are starving because there are no rabbits but that they are going through the experience of starvation with plenty of rabbit meat. For this animal is so lean that illness and death result from being confined to its flesh.

Although both are lean, the meat of hares and rabbits is different. Hare meat is dark while wild rabbit meat is lighter; the meat of domestic rabbits is white, like chicken, to which its flavor is often compared. Recipes for hare and rabbits often suggesting larding the meat with bacon to add flavor, and many involve marinating the meat followed by slow cooking to make it tenderer. Young rabbits and hares are tenderer than older ones. The diets of rabbits and hares reportedly greatly affect the taste of the meat.

Today, Americans eat much less wild hare and wild rabbit meat than do people in Western Europe. Consumption of these wild lagomorphs in Europe is generally limited to those killed by hunters during the hunting seasons, but wild hare meat is also imported from Argentina and Chile, where the introduced European hares are a pest species. Domestic rabbit meat is available at all times, and rabbit farming for meat is a growing industry. A 2003 report from the European Food Safety Authority estimated that about 857 million rabbits were produced for meat worldwide. China produces the lion's share of rabbit meat, followed by Italy and France, while the United States accounts for only about 2 million rabbits per year.

In Europe, hare meat has long been considered superior to rabbit meat, while rabbit meat has had varying status as a high- and low-prestige menu item. During World War II, domestic rabbit meat was promoted as an alternative to scarce beef, but it was neglected after that. Today, in the United States, domestic rabbit is served almost exclusively in high-end restaurants and sold at specialty markets in urban areas. In 2000, the per capita consumption of rabbit meat in the United States was estimated at a mere 0.56 gram (0.02 pound, or about a third of an ounce).

Rabbits and hares are hunted not just for their meat but also for the joy of the chase. Ancient Greek writer Xenophon, who lived from about 430 to 354 BCE, devoted a chapter in his book *On Hunting, a Sportsman's Manual, or, Cynegeticus* [hunting with dogs] to hunting hares with nary a mention of the fruit of the effort. Instead he writes of the hare, "So winsome a creature is it, that to note the whole of the proceedings from the start—the quest by scent, the find, the pack in pursuit full cry, the final capture—a man might well forget all other loves." Xenophon also made astute observations about hare behavior, anticipating much of what scientists have learned about them.

By some accounts, ancient Greeks created the greyhound breed to hunt hares. The greyhound is the only dog whose speed and agility matches that of hares, although greyhounds may have had a much earlier origin as hunters of speedy game; dogs resembling greyhounds are depicted in ruins in Turkey dating to 6000 BCE. Romans are said to have invented the sport of coursing and were explicit about the fact that the goal was to enjoy watching the prowess of the greyhound and hare, not to catch dinner. The Romans also introduced the European hare and greyhounds into Great Britain and, much later, part of the impetus for introducing hares into Australia, New Zealand, and South America was for sport hunting. Throughout Europe, hares remain the most important and popular quarry of sports hunters and much of the pressure for addressing the problem of declining hare numbers comes from sports hunters.

Similarly, the European rabbit was and is popular with sports hunters as well, wherever it occurs. It is by far the most important game species in Spain and Portugal. There are an estimated 1.3 million rabbit hunters in Spain who take about 3 million rabbits per year, and another 300,000 hunters in Portugal who each take at least one rabbit per year. And these numbers are actually lower than they once were because of the serious decline of rabbits on the Iberian Peninsula. For instance, in the 1980s, more than 10 million rabbits were hunted in Spain alone.

In North America, cottontails, jackrabbits, and snowshoe hares are all taken by sports hunters. Cottontails, especially eastern cottontails, are one of the most sought-after game species in the United States, being most popular in parts of the northeast, southeast, and Midwest. One of the challenges facing conservation efforts on behalf of the New England cottontail,

Hunting rabbits and hares for food and sport has a long history. *Der November* by Joachim von Sandrart (1606–1688). Photo from The Yorck Project, Wikimedia Commons / PD

which is easily confused with the eastern cottontail, is the reluctance of states to ban hunting of such an important game species. This is similarly a problem for the vulnerable Corsican hare. On Corsica, it is considered a game species because it is so easily confused with the European hare. In Italy, where it is protected, hunters still mix the two up so protection is hard to enforce.

What products are made from rabbits?

Cry Baby Bunting
Daddy's gone a-hunting
Gone to fetch a rabbit skin
To wrap the Baby Bunting in
Cry Baby Bunting.

After meat for human consumption, fur is the most prominent rabbit product. Undoubtedly, early humans used the fur of the rabbits and hares they ate for clothing, although scientific evidence is lacking because fur doesn't preserve well. Use of rabbit and hare fur for apparel is well documented among Native Americans, however, where pelts were tanned with the fur on or not tanned at all. The *Encyclopedia of American Indian Costume* reports that throughout the continent, rabbit-skin robes were the most common garments after those made of deerskin. These robes were usually fashioned so that both sides were fur covered and as many as 40 skins were required for a single garment. Rabbit fur was used to line and trim moccasins, as headbands and hats, for bags used to carry infants, and by women as menstrual pads.

Rabbit pelts are small and quite fragile compared with those of other mammals skins used for clothing, and rabbit fur has never been as prized as beaver, mink, lynx, and other sought-after furs. But rabbit fur coats are marketed as lower-cost alternatives to other more extravagant fur coats, and the fur may be dyed various colors to appeal to trendy tastes, so the rabbit fur industry is growing. Rabbit fur is also widely used to trim clothing, shoes (especially moccasins) and to decorate toys and crafts; rabbit hairs are also fashioned into fishing flies.

Much rabbit hair was and is turned into felt, a trend that began when beaver, whose fur produces higher-quality felt, nearly disappeared at the end of the nineteenth century after decades of overtrapping in North America. Today, there is a large market for rabbit felt, which is used to produce cowboy hats, especially in Australia, where hat manufacturers import large numbers of farmed rabbit skins from China.

Fur was once primarily a by-product of hunting rabbits and, more recently, of meat production on domestic rabbit farms. But today rabbits farmed for meat are usually slaughtered at 10 to 12 weeks of age, while good quality furs come from adult rabbits and best quality from rabbits in their winter coats. According to a report by the Food and Agriculture Organization of the United Nations, the pelts of rabbits raised for meat are usually simply thrown away, although hair is sometimes used for felt and hides to make fertilizer and glue. Europe has long dominated the rabbit-fur industry, while the United States is a small player. The industry is growing rapidly in China, however, primarily for export.

Fur from the long-haired Angora breed, used primarily to make luxurious sweaters, is in a separate category. Considered a high-value fur for its fineness and luster, it comes from rabbits that are shorn, like sheep, or depilated, at regular intervals, rather than killed.

Animal-rights activists decry the deplorable conditions of rabbits raised in both fur and meat farms. In factory farms, rabbits live in tiny, bare cages, either alone or jammed in with too many others. Caging prevents the rabbits from performing normal behaviors; in some cases rabbits do not even have

The fur of domestic rabbits is fashioned into coats and other garments as an inexpensive alternative to high-end furs like mink and sable.

Photo © Lala Seidensticker

the room to sit up with their ears erect, a characteristic posture. According to the Coalition to Abolish the Fur Trade, mortality rates for farmed rabbits are higher than those of any other animals on commercial farms.

Lagomorphs also figure in folk and traditional medicine. The ancient Romans, for instance, believed eating hare meat would cure sexual problems or would make a man attractive for 9 days. In North America, placing a rabbit's foot under the pillow was supposed to help induce labor in pregnant women. While not strictly a medicinal use, carrying a rabbit's foot was believed to confer good luck in North American folklore, perhaps stemming from African-American folk magic. Similar beliefs about a rabbit's foot were found in many other parts of the world, too. Chinese traditional medicine prescribes eating rabbits (or, presumably, hares) to alleviate fatigue and various digestive problems such as constipation. In India, rabbits (probably Indian hares) have a variety of medicinal uses. Meat is prescribed to treat typhoid and to cure menstrual disorders. Drinking rabbit blood is said to treat asthma; applied externally, the blood is used to heal swellings. The smoke of burning fecal pellets is used to treat hemmorhoids.

Andrew Smith told us that Central Asian folk medicine includes a brew of pika-pellet tea that is prescribed to treat rheumatism. Food writer M. F. K. Fisher, in her 1961 book, *A Cordiall Water: A Garland of Odd and Old Receipts to Assuage the Ills of Man & Beast*, related learning from a Kansas girl the following recipe for reducing a fever: "Gather plenty of turds from the wild jackrabbit, and dry them in the oven to keep for winter in a jar. When the fever will not break, make a very strong tea of the dung and hot water, strain it, and drink it every half hour until the sweating starts. This never fails."

Recently, Chinese investigators of traditional Chinese medicine found that bile from the gall bladder of domestic rabbits was nearly as effective as that of bile from bears in treating a variety of illnesses. This suggested to them that rabbit bile could be a substitute for products from bears, which are being driven to extinction in some places by the demand for medicine from their gall bladders. However, synthetic forms of the active ingredient in bile are available, as are many other scientifically formulated Western medications to treat the same illnesses.

Why do so many rabbits get hit by cars?

Rabbits and hares certainly get killed by cars, as do many other animals. In a 1998 review of the effects of roads on wildlife, landscape ecologists Richard T. T. Forman and Lauren E. Alexander concluded that "sometime during the last three decades, roads with vehicles probably overtook hunting as the leading direct human cause of vertebrate mortality on land." But there is no evidence for or against lagomorphs being killed in collisions with motor vehicles disproportionately to their numbers. No data are collected in a systematic way in the United States on the numbers of small mammals killed on roads, but squirrels, raccoons, Virginia opossums (*Didelphis virginiana*), and, in some areas of the country, armadillos (*Dasypus novemcinctus*), seem to be the most prominent victims.

A survey in Britain reported in 2008 revealed that European rabbits accounted for 58% of road-killed mammals, which the authors attributed to the fact that there are a lot of rabbits there, rather than any special propensity for rabbits to be killed on roads. In fact, they suggested that an individual rabbit may be less likely to be killed than a hedgehog (*Erinaceus europaeus*) because rabbits react to a vehicle's approach when it is about 160 meters (525 feet) away versus just 8 meters (26 feet) away for the hedgehog. However, rabbit and hare escape behavior, which includes freezing followed by running away with erratic changes of direction, may make it more difficult for a motorist to avoid hitting a fleeing lagomorph.

Similarly, in a detailed study in California's Central Valley, black-tailed jackrabbits were the most frequent roadkills, with an average of just over four carcasses per kilometer (about 2.5 per mile) surveyed, with one found dead about every 5 days of the study. Here too, however, this high rate was attributed to the jackrabbit's being the area's most abundant small mammal in the area surveyed.

A questionnaire survey of drivers in Sweden revealed that between 63,500 and 81,500 hares were killed by vehicles in 1992 and that they were the most often killed of the mammals. Again, this is likely due to hares' greater relative abundance, but the authors found that over the past 40 years, the ratio of estimated roadkill to the annual harvest, a measure of abundance, increased twofold, which they attribute to an increase in the volume of traffic over that period.

A study in Alberta, Canada, examined some of the factors contributing to road-killed small mammals, where snowshoe hares accounted for about a quarter of mammals found dead on the road, exceeded in frequency only by red squirrels. The researchers found that snowshoe hares were most likely to be killed where there was vegetation cover close to the road, not surprising given that this species occupies forested habitat and forages in small clearings. Roadsides are often inadvertently attractive to rabbits and hares because they are planted with vegetation that serves as food plants and are mowed to keep vegetation low, which rabbits prefer.

In Spain, roadsides are attractive to European rabbits, which often build their warrens in road ditches, using the road embankments as supporting structures to keep the warrens from collapsing. In a study addressing road-kill mortality in polecats (*Mustela puturius*), which are declining in Spain, researchers found that the presence of rabbit warrens close to roadsides was the most important feature associated with polecat roadkills. The polecats were going where their primary prey was abundant and getting killed while hunting in the warrens or scavenging road-killed rabbits, which were also quite common in the study area.

Roads can have other effects on lagomorphs, however, outside of direct mortality due to vehicle collisions. Roads fragment habitats and avoidance of roads may reduce available habitat. Broom hares in Spain, for instance, prefer habitats with limited human accessibility as measured by distance from roads. A study of European hares in Switzerland found that they avoided roads and preferred large, contiguous habitat areas to small patches among roadways, and thus the density of roads was inversely correlated with hare abundance. How much this and associated roadkill mortalities contribute to the decline of hares in Switzerland and elsewhere in

European hares, while not yet considered threatened, are declining in large part as a result of industrial agriculture; roads are also fragmenting their habitat. Photo © Silviu Petrovan

Europe is unclear however. Moreover, this is little evidence that mortality from roadkills has a major impact on the survival over time of most other lagomorph species compared with other threats.

However, roads are considered an important threat to the endangered Lower Keys marsh rabbit, whose remaining habitat is highly fragmented and whose total population is very small. Traffic interferes with the dispersal of young rabbits away from the site of their birth, with males being particularly susceptible to being road-killed. This may prevent genetic interchange between subpopulations, leading to increased inbreeding and its associated deleterious effects. Moreover, any excess mortality in a species or subspecies existing at such low numbers can be detrimental to the survival of the species. Roads also limit or prevent movement between habitat patches in New England cottontails.

Finally, the presence of roads that make remote areas accessible to people generally increases the incidence of hunting and poaching of all mammalian species and accelerates deforestation and habitat destruction. For threatened lagomorphs subject to hunting and poaching pressures and habitat loss—and that's most of them—roads leading to their habitats are a bad thing.

Rabbits: The Animal Answer Guide

Do house cats kill rabbits?

Free-ranging house cats and feral domestic cats do kill rabbits, which should be expected, given the diet of their wildcat ancestors. African wildcats (*Felis sylvestris*) include scrub hares and rock rabbits in their diets, and European wildcats prefer to hunt European rabbits over rodents wherever they are available. They also take hares, especially young ones as the maximum prey size for a wildcat is about 3 or 4 kilograms (6.6 or 8.8 pounds). Where both have been introduced on islands, European rabbits are the preferred prey of domestic cats, and the cats depress rabbit numbers (see "Are rabbits pests?" in chapter 9). In a long-term study of feral cats and their prey in New Zealand, cats killed almost all young rabbits soon after they emerged from the nest. Rabbits are also taken in substantial numbers by feral cats in Australia.

A study by Great Britain's Mammal Society, in which cat owners were surveyed to document the species brought home by their pets, found that rabbits (and a single hare) formed about 9% of the nearly 10,000 mammals brought home by about 1,000 cats over a 5-month period. Using an estimate of 9 million house and feral cats living in Great Britain, the study authors suggested that cats might account for the deaths of about 92 million wild animals (including mammals, birds, and reptiles and amphibians) in a similar amount of time. Extrapolating from this, that translates into more than about 10 million rabbits a year and a substantial number of hares.

In the United States, there are an estimated 80 million house cats and another 60 to 100 million feral cats. There are no precise figures or breakdowns of prey types, but some experts suggest that these cats kill more than a billion small mammals, including cottontail rabbits and small hares. Here and elsewhere, cats vastly outnumber wild medium-sized predators and may compete with bobcats, foxes, raptors, and other carnivores for rabbit prey.

Predation by cats is cited as risk factor contributing to the endangerment of Lower Keys marsh rabbits in south Florida, riparian brush rabbits in California, the riverine rabbit in South Africa, and the Amami rabbit in Japan.

What can an ordinary citizen do to help rabbits?

You yourself, don't you find it a beautiful clean thought, a world empty of people, just uninterrupted grass, and a hare sitting up?

D. H. Lawrence

The greatest threat to rabbits and all of the world's wildlife is the enormous human impact on the landscape. A recent detailed analysis showed people have some direct influence on 83% of the Earth's land surface, ranging from living on the land and using it for agriculture and other human activities to having access to land via roads and other transportation corridors to subjecting it to various forms of pollution. A measure of an individual's impact on the environment is called an "ecological footprint" and takes into account such factors as land, energy, and water usage, as well as pollution production. By some estimates, there about 1.4 hectares (4.5 acres) of habitable land for each person, but the ecological footprint of many people, especially in rich Western nations, is already far greater than this. The aspirations of people in less-rich nations for a Western lifestyle means that their current relatively small footprint will grow larger, as it is doing in India and China. Some experts fear that without changes in our unsustainable consumption of natural resources, the future looks bleak for people and wildlife.

It will take the concerted action of governments and corporations around the globe to address these issues, for instance by coming together to reduce the carbon emissions that contribute to global warming and develop clean, renewable sources of energy, but every individual can make changes in everyday activities to reduce their ecological footprint. The list of things to do is familiar to everybody—reuse, recycle, reduce energy consumption—but everybody needs to commit to doing them. Everyone can use their purchasing power to reward environmentally friendly corporations, vote for political leaders who are working toward a more sustainable future, and volunteer at or contribute to organizations working on behalf of the environment.

More specifically related to lagomorphs, people can directly reduce some threats to their survival, such as by avoiding the use of potentially harmful herbicides, pesticides, and rodenticides and by keeping their domestic cats indoors and having them spayed or neutered.

Individuals can also learn about lagomorphs and share their knowledge with family, friends, and neighbors. Lagomorph conservationists note that one of their greatest challenges is the widespread belief that "rabbits breed like rabbits," which makes it hard to get the message out that many lagomorphs species are endangered and most are declining. Lack of awareness means that mustering political support for governmental funding of lagomorph conservation is challenging, as is raising the funds from individuals and corporations that supplement most conservation programs for endangered species. Every citizen can become an ambassador for bunnies!

In some cases, private landowners can contribute by maintaining or creating rabbit-friendly landscapes on their property. In South Africa, all the habitat of the critically endangered riverine rabbit is on private farmland. Conservationists there are working with owners to encourage them to protect critical habitat on their land and to manage fires as well as farming and grazing regimes in such a way as to not harm the ecosystem. Similarly, the future of the endangered New England cottontail depends on private landowners agreeing to create and manage the brushy habitat this species requires. Experts estimate that creating a mere 809 hectares (2,000 acres)—just over 8 square kilometers (about 3 square miles)—of high-quality new and restored habitat in New Hampshire and Maine and protecting an additional 2 square kilometers (500 acres) could result in 11 to 21% percent annual increases in the numbers of this unique cottontail over 10 years, instead of recent declines of 5 to 9% annually.

Chapter 11

Rabbits in Stories and Literature

What roles do rabbits play in mythology and religion?

Rabbits and hares may well be the animals most represented in the myths, fables, folktales, symbols, and religions of people around the world, playing prominent and often surprisingly similar roles in the diverse cultures of Europe, Asia, North and Central America, and Africa. Yet there is little congruence in the various roles that rabbits play—they can be depicted as benign, gentle souls as well as monsters. Rabbits—and we will use rabbits here to mean both rabbits and hares—symbolize timidity and fearfulness, as well as lustful female sexuality. Rabbits are almost ubiquitously depicted as tricksters, using cunning and guile to best creatures that want to eat them, yet they also act as metaphors for shyness. Rabbits often feature as the familiars of witches but also appear in paintings with the virginal Madonna. Rabbits may mean good luck, or bad. They may be a link to the heavens or to hell. Susan E. Davis and Margo DeMello did a masterful job of reviewing rabbits as cultural icons in *Stories Rabbits Tell: A Natural and Cultural History of a Misunderstood Creature*, and a great deal of the following is based on their work.

In mythology and ancient religions, rabbits are often associated with feminine principles and with the moon, which is also usually seen as feminine, with the moon's waxing and waning symbolizing rebirth as well as fertility and menstruation. But as if to emphasize the paradoxical place of rabbits in the human psyche, in Egyptian mythology, the god Osiris, who was associated with the sun and the moon and also represented rebirth, was sometimes pictured with a rabbit's head; in the form of a rabbit. Osiris was sacrificed to the Nile to ensure its annual flooding, which was essential to agriculture.

In cultures as diverse as China and Maya Mexico, people venerated moon goddesses depicted as or associated with rabbits. Among the Maya, the moon goddess who also oversees childbirth is portrayed holding a rabbit. In China, a moon goddess named Gwallen is represented by a rabbit, which is a yin, or feminine, animal. A moon goddess worshiped by the Ugric people of western Siberia was manifested as a rabbit. Freyja, a Norse goddess of love, sensuality, and women's mysteries, was attended by rabbits. Because they represented fertility, rabbits were sacred to Aphrodite, Greek goddess of love and marriage.

In Anglo-Saxon myth, Ostara is the goddess of the moon, fertility, and spring—a period when life is reborn after northern Europe's barren winter months. Ostara's familiar animal was a rabbit, symbol of fertility. Ostara's Celtic counterpart, Eostre, turned into a rabbit at each full moon. These goddesses also represented female sexuality, as rabbits continue to do in contemporary society. Linguistically, from Ostara or Eostre comes the words "estrogen," the female sex hormone, and "estrus," the technical term for heat. It is also the origin of the word "Easter" and of the Easter Bunny, who delivers and hides colored eggs to children as part of the Christian celebration of the resurrection of Jesus Christ.

In Anglo-Saxon tradition, a festival was held in Ostara's honor in April, when fires were lit at dawn to protect crops from frost. In some Anglo-Saxon and German dialects, April is called Ostara's month. According to one story, Ostara transformed a pet bird into a rabbit to entertain some children, and the rabbit proceeded to lay colored eggs that the goddess then gave to the little ones. In another version, a small girl asked the goddess to save a bird she found nearly dead from the cold. The goddess saved the bird by turning it into a rabbit, which produced colored eggs. Like rabbits, eggs are symbols of fertility.

When Anglo-Saxons converted to Christianity about 550 CE, the celebration of the resurrection of Christ was grafted onto and replaced the pagan festival—both, after all, were dedicated to the idea of rebirth—and the name survived. In fact, English and German are among the few languages in which the word for Easter is still related to the goddess Eostre; in most other European languages, the name is a derivation from the Hebrew pasach, or Passover, suggesting the day's link to the Jewish holiday.

Some of the old symbols, such as bunnies and eggs, were brought to the United States by German immigrants. And with the American genius for gussying up any event with presents, it wasn't long before the Easter Bunny was delivering chocolate coast to coast.

In many cultures, people see a rabbit in the moon, not a man in the moon. *Songoku and jewel hare* by Yoshitoshi Taiso (1839–1892). Photo from Library of Congress, Prints & Photographs Division, LC-DIG-jpd-00825

And it's hardly a leap, or even a hop, from Ostara the sex goddess and her totemic rabbit to the Playboy Bunny. If evolutionary biologists are right, sexual attractiveness is really all about fertility, the goal of all right-thinking men is to produce as many children as possible. Then, of course, a rabbit makes a pretty sexy symbol.

In Chinese, Japanese, Indian, and other Asian cultures, rabbits take on a different association with the moon than we are familiar with in the West. In these cultures, what Westerners call "the man in the moon" is called "the rabbit in the moon." According to one Chinese legend, which is a variant of a similar Indian tale, the Buddha sent the rabbit to the moon, rewarding it with immortality in gratitude for offering itself as food to the hungry monk. In some Japanese images, the rabbit in the moon is shown with a mortar and pestle and is said to be mixing a potion that confers immortality. Ancient Mexican cultures also talk of the rabbit in the moon. Surprisingly, an Aztec legend tells a very similar story of rabbit self-sacrifice being rewarded with immortality on the moon, with the Buddha role played by the god Quetzalcoatl. In a Siberian legend, a conclave of animals sends a rabbit to recapture the sun from evil spirits who had stolen it. With evil spirits in hot pursuit, the running rabbit kicked a ball of fire with his hind legs, breaking it into two parts. The rabbit then kicked the smaller part into sky to become the moon and larger to become the sun.

Oddly enough, the rabbit in the moon was the subject of an exchange between the Mission Control Center in Houston, Texas, and Apollo 11 astronaut Michael Collins in the 1969 mission to land on the moon:

HOUSTON: Among the large headlines concerning Apollo this morning there's one asking that you watch for a lovely girl with a big rabbit. An ancient legend says a beautiful Chinese girl called Chang-o has been living there for 4,000 years. It seems she was banished to the moon because she stole the pill for immortality from her husband. You might also look for her companion, a large Chinese rabbit, who is easy to spot since he is only standing on his hind feet in the shade of a cinnamon tree. The name of the rabbit is not recorded.

COLLINS: Okay, we'll keep a close eye for the bunny girl.

Artemis and Diana, the Greek and Roman goddesses of the hunt, wild places, and protectors of wild animals, were also associated with rabbits but, again paradoxically, they were known for their chastity. Ancient Greek hunters were instructed by Artemis to leave newborn rabbits to her protection. In his hunting guide (see "Why do people hunt rabbits?" in chapter 10), Xenophon echoed this, writing, "Every true sportsman, however, will leave these quite young creatures to roam freely. 'They are for the goddess.'" In the myths of some North America natives, rabbits are credited with teaching people to hunt and fish as well as with creating the Earth.

The constellation Lepus was named by the astronomer Ptolemy about 150 CE, but myths about Lepus are known from ancient Greece. In one, Lepus, which appears to be the feet of Orion, was placed in the sky to benefit the great hunter. The rabbit is also both one of the 12 Chinese signs of the zodiac and a day name in the calendar of the Aztecs. The Navajo called the stars in the tail of the Scorpio constellation the Rabbit Tracks, and in their tradition this constellation governed hunting. Its position in the sky dictated the beginning of the hunting season in the late fall and its ending in the spring.

There is a mysterious symbol of three hares chasing themselves in a never-ending circle in which each individual rabbit has two ears, but the three rabbits together have a total of only three ears. Its meaning is unknown, but it appears in Buddhist cave temples in China dating to the Sui Dynasty (581–618 CE), in a copper coin minted in Islamic Iran in the 1200s, in churches and homes in western Europe and especially in Devon in England from the seventeenth century, and in an eighteenth-century Jewish synagogue in Germany. Scholars puzzle over the meaning of symbol, which they believe traveled from China through the Middle East to Europe via the Silk Road.

What roles have rabbits played in language and literature?

Among the best known of Aesop's fables is "The Tortoise and Hare," but hares figure in many of his other fables as well. These tales of anthropomorphized animals were meant and still serve today as moral lessons, such as "slow and steady wins the race." In "Jupiter and the Hare," the hare is unhappy with being a weak, insubstantial creature and asks Jupiter to give him great antlers like those of a deer so that other animals would fear him. Jupiter tells the hare he won't be able to carry the weight of antlers, but hare insists he can and so Jupiter crowns him with large, branching antlers. Sure enough, the antlered hare is unable to run and is killed by a shepherd. The moral, people often wish for what will bring them prestige but will result in their death. (We can't help but wonder whether the mythical "jackalope" of the western United States—a cross between a jackrabbit and an antelope that bears horns—has its roots in this fable.)

In another fable, "The Hares and the Frog," hares are so ashamed of their timidity that they decide to commit suicide. Approaching a pond to drown themselves, the hares startle the frogs basking on the banks and the frogs leap into the water. Seeing this, the hares abandon their suicidal intentions because they realize that there are creatures even more cowardly than they are. The moral is, "unhappy people are comforted by the sight of someone who is worse off than they are," a sentiment summarized in the German word *Schadenfreude*.

For the Greeks and in Western literature ever since, hares were the proverbial cowards; "a hare's life" referred to someone who was constantly fearful. But timid hares also appear in other cultures; for instance, in a story from India, a hare plays the role of Chicken Little. Using hares and rabbits as symbols of timid people is a standard literary device. We speak of someone looking like a "scared rabbit" and a person who is shy and diffident is described as "rabbity." In the play *King John*, Shakespeare uses the hare's well-known timidity to tweak the cowardly king:

You are the hare of whom the proverb goes,
Whose valour plucks dead lions by the beard.

Among native North Americans, the rabbit is more often a trickster figure, another nearly ubiquitous rabbit motif in which a rabbit uses both cleverness and deceit to get out of trouble, usually to escape being eaten by predator. Rabbit tricksters also appear in Asian and African folktales. African slaves brought their trickster tales to the Unites States, where they were later collected by Joel Chandler Harris in stories of Br'er Rabbit, as narrated by his Uncle Remus character. The cartoon character Bugs Bunny is a modern take on the trickster rabbit theme: he invariably outwits his gun-toting nemesis Elmer Fudd. Some speculate that the rabbit as trickster represents oppressed people—the prey—who take comfort in the idea of overcoming their oppressors—the predators.

During the Soviet era in Russia, a rabbit was used in a joke that subversively criticized Stalin's show trials of his enemies: "A rabbit was running for his life through the desert. When asked by another rabbit what he was running from, he said a lion had threatened to eat any camel he met. 'But we're rabbits!' 'Yes, but just try to prove that you're not a camel!'"

Rabbits appear in countless children's stories, including such enduring classics as Beatrix Potter's *The Tale of Peter Rabbit*, Margaret Wise Brown's *Runaway Bunny* and *Goodnight Moon*, and Margery Williams's *Velveteen Rabbit: Or How Toys Become Real*. In these stories, the rabbit is a stand-in for a real child; although Peter Rabbit acts in some ways like a rabbit, stealing vegetables from Mr. McGregor's garden, he is really a naughty little boy in rabbit's clothing.

The dithering White Rabbit of Lewis Carroll's *Alice's Adventures in Wonderland* is timid and fearful, dashing into a hole at any hint of trouble; he was "mad as a March hare." The timidity of hares may also be the origin of the phrase "hare-brained" to describe foolish behavior. In a wonderful use of this metaphor, Benjamin Disraeli spoke of "the hare-brained chatter of irresponsible frivolity."

Rabbits and hares were reputedly the familiars of witches, as suggested in this children's poem by Walter de la Mare (1873–1956):

In the black furrow of a field
I saw an old witch-hare this night;
And she cocked a lissome ear,
And she eyed the moon so bright,

And she nibbled of the green;
And I whispered "Wh-s-st! witch-hare,"
Away like a ghostie o'er the field
She fled, and left the moonlight there.

There are a host of sayings that refer to rabbits' reputation for fertility, including the oft-repeated remark about a woman with many children, "who breeds like a rabbit." The word "breeds" is sometimes replaced with a vulgarism for sexual intercourse, used to call a woman promiscuous. Bunny may also describe a promiscuous or sexy woman, most famously in popular culture represented by the Playboy Bunny created by Hugh Hefner. (Incidentally, Hugh Hefner funded the research on the Florida Keys marsh rabbit that helped identify it as a distinct subspecies; *Sylvilagus palustris hefneri* was named for him.) On the other hand, bunny may be applied to a woman or a child as a term of affection, as in "honey-bunny" and "cute as a bunny."

The best-selling *Tales from Watership Down* by Richard Adams is a novel for children and adults in which, except for the fact that they can talk and have culture, the rabbits generally behave like rabbits. Adams' used as his model the nonfiction *The Private Life of the Rabbit* by British naturalist Ronald Lockley. In this story, rabbits in an English warren get wind of a land developer's plan to destroy their habitat. In their search to find and establish a new home, they face constant danger and suffer the cruelty of a rabbit tyrant who must be defeated before the other rabbits can thrive in their new home. *Watership Down* is a great adventure story, but it is also viewed as an allegory of the human struggle between tyranny and freedom or between the individual and the corporate state.

In *Henry V*, Shakespeare metaphorically employed the thrill of hunting hares with hounds, in the king's stirring speech to rally his troops to battle:

I see you stand like greyhounds in the slips,
Straining upon the start. The game's afoot:
Follow your spirit, and upon this charge
Cry "God for Harry, England, and Saint George!"

The game, of course, refers to the hare. "Come, Watson, come. The game is afoot" is also Sherlock Holmes's rallying cry to Dr. Watson in the opening of "The Adventure of the Abbey Grange"; Arthur Conan Doyle was undoubtedly quoting Shakespeare.

In Shakespeare's *As You Like It*, Rosalind uses a hare-hunting metaphor, saying, "Her love is not the hare that I do hunt."

French philosopher Blaise Pascal (1623–1662) also used the familiar

Cute as a bunny. Photo © Susan Lumpkin

sport of hunting hares to describe the human motivation for striving for success: "We like the chase better than the quarry . . . And those who philosophize on the matter, and who think men unreasonable for spending a whole day in chasing a hare which they would not have bought, scarce know our nature. The hare in itself would not screen us from the sight of death and calamities; but the chase, which turns away our attention from these, does screen us."

A proverb stems from a sixteenth-century writer: "There is no [more] suche tytifils [scoundrels] in Englands grounde, To holde with the hare, and run with the hounde." Now most often expressed as "You cannot run with the hare and hunt with the hounds," it refers to people who behave hypocritically.

In a haunting metaphor for the pain of great loss, British poet Dame Edith Sitwell (1887–1964) wrote in her poem "Still Falls the Rain" of the "self-murdered heart" and of

The wounds of the baited bear,—
The blind and weeping bear whom the keepers beat
On his helpless flesh . . . the tears of the hunted hare.

Rabbits in Stories and Literature

Another lovely metaphor comes from the great Arabic writer Al Jahiz (771–868): "The duration of life is but the hop of a hare."

Sometimes, the rabbit represents itself, as in this verse from "Love on the Farm" by D. H. Lawrence (1885–1930):

> The rabbit presses back her ears,
> Turns back her liquid, anguished eyes
> And crouches low: then with wild spring
> Spurts from the terror of *his* oncoming
> To be choked back, the wire ring
> Her frantic effort throttling:
> Piteous brown ball of quivering fears!

And in this description from John Steinbeck's *Of Mice and Men*: "On the sand banks the rabbits sat as quietly as little gray sculptured stones."

What roles do rabbits play in popular culture?

Apart from their role in popular literature for children and adults, rabbits appear amazingly often in popular culture once you start looking. Endearing little bunny rabbits appear on children's blankets and clothing and are very popular as plush toys. Hundreds of children's products are spin-offs of the immensely popular *The Tale of Peter Rabbit*. Bugs Bunny is just one of dozens of cartoon rabbits that populate the fantasy world of children. Rabbits are featured in countless kitschy collectibles and decorative items for the home and garden.

Rabbits are used to market products as diverse as the Energizer Bunny (batteries); Trix (children's cereal); Reader Rabbit (educational software); and Volkswagen Rabbits (cars). This last is a rare exception to the use of more powerful animal images to market cars, such as Cougar, Stingray, and Thunderbird, although with the rabbit's reputation for speed, possibly a step up from the Volkswagen Beetle. Annie's Homegrown, a line of organic food products that includes Cheddar Bunnies and Bunny Grahams, uses a rabbit in its logo, as does PETA (People for the Ethical Treatment of Animals) and, appropriately, the Coalition for Consumer Information on Cosmetics. French Rabbit wines are touted as coming from grapes "sourced from sustainably farmed vineyards" in southern France. The Duck-Rabbit Craft Brewery makes beer and uses as its logo the famous duck-rabbit visual illusion. The mascot of South Dakota State University's sports teams is a jackrabbit.

We have described television antennas as "rabbit ears" and salad as "rabbit food." Magicians pull rabbits out of their hats, and the phrase also describes a person's accomplishing something that is seemingly impossible.

This sign welcomes guests to the Rabbit Resort in Pattaya, Thailand.
Photo © John Seidensticker

Rabbit kitsch in the form of salt-and-pepper shakers. Photo ©
John Seidensticker

In a twist on the metaphorically timid rabbit, there are several examples of monster or killer rabbits in films, whose humor comes from the unlikelihood of such a creature. *Night of the Lepus* is a 1972 mock-horror movie in which "husband-and-wife scientists unwittingly unleash a horde of giant man-eating rabbits," as summarized in the TCM Movie Database. Interestingly, the preposterous plot has a ring of ecological reality. The scientists are testing a serum to control the rabbits whose population exploded

French rabbit

Rabbits and hares are featured in the logo of many products, including French Rabbit wine (*top*) and Annie's Homegrown foods (*bottom*).

Courtesy of French Rabbit, Boisset America, and Annie's Homegrown

As you read this book, you're more likely to see this image as a rabbit than as a duck; reading a book about ducks, the opposite would be true.

Image from Wikimedia Commons / PD

after coyotes were killed off in Arizona; rather than killing the rabbits, it turned them into super-sized killers who terrorize, attack, and kill people. Finally, they are herded with vehicles to a train track that was wired to electrocute them. In the end, order is restored—but only with the return of the coyote.

In the 1975 film, *Monty Python and the Holy Grail*, the Knights of the Round Table must overcome the vicious Rabbit of Caerbannog, which appears to be a harmless white rabbit until it decapitates the first knight sent to kill it. Even when the knights attack in force, the rabbit manages to kill two of them as the others flee shouting, "Run away, run away!"

The popular 1979 children's book, *Bunnicula: A Rabbit-Tale of Mystery* by Deborah and James Howe, is the first of a series of books told from the point of view of a pet dog named Harold. Harold is persuaded by a cat named Chester that the rabbit their family brought home is a vampire. When a white tomato appears in the kitchen, Chester is sure it's been sucked dry by the vampire bunny and ominously warns Harold, "Today vegetables. Tomorrow . . . the world!" Hilarity ensues as Chester tries to alert the family of the danger they are in and is constantly misunderstood, but the funniest scenes are early on, as Chester amasses the evidence that Bunnicula is a blood-sucking villain.

Brilliant humorist P. G. Wodehouse in his 1901 story, *The Swoop, or How Clarence Saved England*, memorably wrote of the heroic Clarence, "He could do everything that a Boy Scout must learn to do . . . He could imitate the cry of a turnip in order to deceive rabbits."

Chapter 12

"Rabbitology"

Who studies rabbits?

A variety of specialists study lagomorphs. Scientists who study pikas, rabbits, and hares include wildlife biologists and ecologists, who generally observe the lagomorphs in their natural environments to better understand how these animals behave, reproduce, and fulfill various roles in the ecosystems of which they are a part and what they need to survive in our changing world. Paleontologists pore over the fossils of extinct lagomorphs to learn about their evolution, diversification, and geographic distribution over time. Comparative anatomists and molecular biologists tackle the thorny problem of sorting out how many species of lagomorphs exist and how they are related to one another. Anthropologists examine the roles of lagomorphs in the diets and cultures of early and modern humans.

In addition, the best-known species, the European rabbit in both its wild and domestic form, is studied by these kinds of scientists as well as by veterinarians, biomedical researchers, agricultural scientists, geneticists, and many more. They are studied in laboratories, in the field, and in-between in large enclosures that mimic the natural environment. European rabbits are studied to learn more about them in their own right and also to serve as animal models for exploring physiological processes relevant to understanding and treating human diseases. Many scientists are working to improve domestic rabbit breeds so they produce more meat or fur more efficiently.

Scientists concerned primarily with domestic food and fur rabbits are often members of the World Rabbit Science Association (http://world-rabbit-science.com/). Some lagomorph scientists serve on the Lagomorph

Specialist Group of the International Union for the Conservation of Nature, or IUCN. This group conducts thorough evaluations of the conservation status of lagomorphs to determine how species should be rated on the Red List of Threatened Species. The World Lagomorph Society is a relatively new organization with the goal of improving communication among scientists interested in lagomorph biology and conservation; nonspecialists with an interest in rabbits are welcome to join this society too.

What seems clear, however, is that there aren't enough scientists studying lagomorphs! With a handful of notable exceptions, species such as the snowshoe hare, some populations of European hares and mountain hares, and few pikas, and topics such as fossil history and molecular biology, wild lagomorphs are relatively neglected by scientists.

Which species are best known?

The European rabbit wins the best-known contest, hands down, while there are many contenders for the title of least known. As a rough comparative measure, we searched the ISA Web of Science database, which indexes the articles in many scientific journals, using the scientific names of the various species as key words. The European rabbit search yielded 1,732 titles for this single species, compared with 1,193 for all the hares combined, 311 for all the cottontails, 78 for all of the other rabbits, and 283 for all of the pikas.

In part, this disparity stems from the extensive use of domestic European rabbits in biomedical research, where the goal is not to understand the species per se but rather to acquire knowledge about basic physiological and other processes, especially for application to human health. This is the expressed goal of efforts to sequence the rabbit genome, as noted on the U.S. National Center for Biotechnology Information's Web site: "The rabbit was selected by the National Human Genome Research Institute for whole genome sequencing to enhance our understanding of the functional and structural elements of the human genome and to markedly increase the rabbit's experimental value as an animal model for human disease."

As noted earlier, there is also a vigorous effort to study domestic rabbits in production systems, as a source of food and, to a lesser extent, fur. But this species has also demanded attention because of its importance as a game species and as an invasive species throughout the world. On the Iberian Peninsula, rabbits are the premier game species. With their decline, much research in Spain and Portugal is focused on recovering rabbit numbers for the sake of the rabbits as well as for human hunters and to help Iberian lynx recover from near extinction. In contrast, Australian biologists, in particular, have developed a huge body of knowledge about

European rabbits as part of efforts to control their numbers. British and German biologists have conducted long-term studies of European rabbits in large, seminatural enclosures to describe the intricacies of their social behavior.

European rabbits are also relatively easy to study compared with many other lagomorphs and many other mammals. Their presence is made conspicuous by their extensive burrow systems from which they can readily be captured. They live in relatively open habitats, are large enough to be easily observed, and occur in places that, from the scientists' point of view, are easy to get to and environmentally benign, compared, for example, with remote, cold mountains occupied by most pikas or the dense tropical forest of the striped rabbits.

Dozens and dozens of popular books have been written about domestic European rabbits, aimed primarily at the people who keep or breed them as pets. *Stories Rabbits Tell: A Natural and Cultural History of a Misunderstood Creature* by Susan E. Davis and Margo DeMello, mentioned in chapter 11, takes a more scholarly approach, reviewing the natural history of wild European rabbits, the behavior and care of domestic pet rabbits, rabbit's roles in ancient and modern cultures, and their current uses for food, fur, and scientific research. *The Textbook of Rabbit Medicine* by Frances Harcourt-Brown is a detailed review of the veterinary care of domestic rabbits, and *The Biology of the Laboratory Rabbit*, edited by Patrick J. Manning, Daniel H. Ringler, and Christine E. Newcomer, is a classic text for veterinarians and biomedical scientists. *The European Rabbit: The History and Biology of a Successful Colonizer*, edited by Harry V. Thompson and Carolyn M. King, and published in 1994, is the most recent book to review what was known about the wild form's natural history.

Among the hares, the snowshoe hares have been very well studied, particularly by scientists trying to determine the drivers of the lynx-snowshoe hare cycle. Most notably this research has been conducted by Charles J. Krebs and his students and colleagues but many others are engaged in it as well. The European and mountain hares, important game species in Europe, are reasonably well-known, too, with many studies of their ecology and behavior. Irish biologists have conducted wide-ranging studies of the Irish hare, a subspecies of the mountain hare, to determine the reasons for its decline. A fair amount is also known about the black-tailed jackrabbit.

Among cottontails, only the eastern cottontail has been extensively studied by a variety of scientists, although increasing effort is being devoted to the rare and imperiled New England cottontail, the Lower Keys marsh rabbit, and the pygmy rabbit.

There are an impressive number of studies of American pikas, plateau pikas, and collared pikas, making these the best most known among this

The snowshoe hare is one of the few lagomorphs that have been fairly well studied by scientists. Photo by Betsy L. Howell, USDA Forest Service

group. Relatively little to essentially no research has been done on the rest of the pikas.

Which species are least known?

The simple answer is most of them. As noted above, not much is known about most of the pikas, for instance, or about most of the hares, South American cottontails, or any of the non-cottontail rabbits other than the European rabbit. Even most of the North American cottontails are little known beyond basic natural history.

A few species, however, are exceptionally poorly known. Mexico's Omilteme rabbit is known only from three museum specimens collected in the early part of the last century and one skin recovered from a hunter in 1998. The Annamite and Sumatran striped rabbits are almost equally obscure, with nothing known about their life histories. Almost nothing is known about the ecology of the Bunyoro rabbit of central Africa, and little more about the African rock rabbits or that continent's hares other than the Cape hare.

What is known about wild lagomorphs has been reviewed in a few books, most notably in the 1990 volume *Rabbits, Hares and Pikas: Status Survey and Conservation Action Plan*, edited by Joseph A. Chapman and John E. C. Flux, and the more recent *Lagomorph Biology: Evolution, Ecology, and Conservation*, edited by Paulo Célio Alves, Nuno Ferrand, and Klaus Hackländer, published in 2008. The lagomorph species accounts prepared for the 2008 Red List mammal assessment provide very useful, updated overviews of the biology and conservation of the lagomorphs and are available online at the IUCN's Web site.

Like most African hares and rabbits, little is known about the scrub hare. Photo by Lee R. Berger, Wikimedia Commons / CC-BY-SA 3.0

Anyone wishing to study lagomorphs has plenty of choices of species we need to know more about and of questions to ask. Many of these have come up elsewhere in this book. How many more species of lagomorphs are there? Why do lagomorphs display such a narrow range of sizes compared with rodents and others? Why do many rabbits and hare have such restricted ranges, while others are widespread and adapted to a variety of habitats? How will lagomorphs in different parts of the world deal with the effects of climate change? What environmental conditions favor the survival of rabbits and how can these be preserved, enhanced, or restored to prevent various species' declines and extinctions?

How do scientists tell rabbits apart?

To understand the ecology and life history of lagomorphs it is often essential to be able to identify individual animals. This is the only way, for instance, to follow the fate of individuals over time and to analyze their social interactions. In some mammals, it is possible to identify individuals by their unique markings. The stripe patterns of tigers, for instance, are like fingerprints, and scientists take their photographs so individuals can be recognized. Similarly, elephants can be identified as individuals by unique scars, torn ears, and other variations. Some researchers have used this method to study lagomorphs, recognizing individuals by scars on the ears and unique fur colors and patterns. More often, scientists must capture and mark lagomorphs in some way, either temporarily or permanently.

Mammalogists use a variety of techniques to mark lagomorphs for individual identification at a distance. They may dye the animal's fur or shave

Rabbits: The Animal Answer Guide

a unique pattern into the hair on the back, for instance, although this is temporary and the mark lasts only until the next molt. Metal or plastic tags attached to the ear, in unique combinations of colors, are also used to mark lagomorphs, the size of the tag depending on the size of species being studied. Depending on the goals of the study and whether the research plan includes recapturing the animal, it may also be permanently tattooed on the inside of the ear.

Many studies involve radio-tracking individuals to follow their movements, which is especially useful when the animals are secretive and nocturnal. A small radio transmitter that sends signals to a researcher's receiver is affixed to the animal, usually in collar around its neck. For smaller animals, such as pygmy rabbits, the transmitter may also be glued to animal's back. Radio-tracking devices are often equipped with mortality sensors, altering researchers if the animal hasn't moved in a specified period of time and can be assumed to be dead. To recover the transmitter, the animals may need to be recaptured but many collars are designed to fall off.

Among domestic rabbits, which don't need to be identified at a distance because they can be handled, the most frequent marking method is a tattoo of numbers or letters on the inside of the ear. The rules of the American Rabbit Breeders Association dictate that every rabbit exhibited in a show be tattooed in this fashion. Sometimes veterinarians tattoo the abdomen after a rabbit is spayed or neutered in case there is any doubt in the future that the procedure has been performed.

How can I become an expert on rabbits?

In every generation and among every nation, there are a few individuals with the desire to study the workings of nature; if they did not exist, those nations would perish.

AL JAHIZ

People who are experts on lagomorphs are biologists. Ernst Mayr, an eminent Harvard biologist, said, "Being a biologist does not mean having a job. It means choosing a way of life." People who chose that way of life, first and foremost, are insatiably curious about how the natural world works. Lagomorph biologists, for whatever reasons, have focused their thirst for understanding nature on this group of unusual mammals.

If you want to become a biologist, the first thing to do is go outdoors and look around. Wherever you live—in the depths of a city or near a wilderness—you will find an abundance of biological phenomena to observe and ask questions about. (Research also suggests that spending time outside is important to your physical and mental health! Scientists have identified a

Scientists attach radio-collars to lagomorphs, such as this Irish hare (*left*) and pygmy rabbit (*right*) to track their movements. Irish hare photo by Alan Wolfe, Wikimedia Commons / CC-BY-A 3.0; pygmy rabbit photo © Jim Witham

syndrome associated with lack of outdoor activities. Called "nature deficit disorder," it is linked to childhood obesity, depression, and attention deficit hyperactivity disorder). Biology is concerned with the discovering the answers to three broad questions: What? How? Why? Learning to ask these questions is the first step toward being a biologist.

You can pick any living thing in the woods or in your backyard and ask yourself, what is it? A plant or an animal? What kind of animal: insect, bird, mammal, or what? Then delve more deeply. What kind of bird is it, where and when do you see it, and so on. These are the "what" questions and, as you've learned in this book, there are still plenty of mysteries even about the different kinds of lagomorphs. "How" questions address how things work. How does a bird fly or a vine twist around a tree limb or a rabbit see in the dark? Finally, "why" questions deal with the environmental, geographical, historical, and evolutionary factors that account for everything that exists in nature. Ask yourself, why did moths and bats evolve to fly only at night, why are there no jackrabbits foraging in your urban backyard or no polar bears hunting in Florida? Whatever your questions, try to guess the answer and then do research in books or online to find out whether your idea corresponds to what others have already discovered. This is what scientists do: observe, form hypotheses based on their existing knowledge, and then test them through research, which might involve more observation, formal experiments, or a review of the scientific literature. But in getting started, it's more important for you to develop the habit of curiosity, of asking questions, than to immediately find the answers.

Along with spending time outdoors, read voraciously. In our view, a budding biologist can do nothing more important than reading the works

Rabbits: The Animal Answer Guide

of Charles Darwin, father of the theory of evolution through natural selection. *The Voyage of the Beagle* is a great start, and *On the Origin of Species* is essential. Virtually all of modern biology is basically concerned with elucidating the details of Darwin's theory. Darwin wrote for an educated general audience, so these books are eminently readable and enjoyable.

If you want to become a lagomorph biologist in particular, you've already gotten a start in reading this book. Add to your knowledge by exploring the subjects that you found most interesting; this will help guide you in selecting which area of lagomorph biology you want to specialize in later. If you have the time and inclination, caring for a pet domestic rabbit would be a good way to get familiar with lagomorph behavior. Go searching for wild lagomorphs, too.

The first step in becoming a professional lagomorph biologist, which usually requires at least a master's degree and more often a Ph.D., is to acquire a solid education, heavy on sciences, in high school and college. If you're still in high school, consider attending a university where lagomorph biologists are on staff. This is not essential but can give you a head start on your career—there are often opportunities for undergraduates to assist in research. High school students can also find volunteer opportunities working with lagomorphs, for example with domestic rabbits in a local zoo or with a veterinarian who treats rabbits. For college students, selecting a graduate school and a graduate advisor whose interests match yours is critical. At any stage, using the Internet makes finding the right schools and professors relatively easy. Start by checking out the university affiliations and interests of members of the Lagomorph Specialist Group and the World Lagomorph Society.

If reading about rabbits has inspired you, but you already have a different career, remember that amateurs can often make significant contributions to some aspects of biology. The observations of amateur bird watchers, for instance, have been critical to tracking the movements of migratory birds and changes in birds' distribution and abundance over time. Other "citizen scientists" are helping scientists track monarch butterfly migration, firefly distribution, plant phenology, and much more. With so little known about most lagomorphs, including such basic information as the extent of their distributions, careful, systematic observations by knowledgeable "bunny watchers" of the wild lagomorphs in a nearby woods or field could provide useful information. A lagomorph biologist at a local university or with your area's wildlife department might welcome your volunteer help, too.

Appendix

Rabbits of the World

Scientific Name	Common Name	General Location
Order Lagomorpha		
Family Ochotonidae		
Ochotona alpina	alpine pika	Central Asia
Ochotona argentata	silver pika	China
Ochotona cansus	Gansu pika	China
Ochotona collaris	collared pika	Alaska, Western Canada
Ochotona curzoniae	plateau pika	Western China, Northern South Asia
Ochotona daurica	Daurian pika	Central Asia
Ochotona erythrotis	Chinese red pika	China
Ochotona forresti	Forrest's pika	Northern South Asia
Ochotona gaoligongensis	Gaoligong pika	China
Ochotona gloveri	Glover's pika	China
Ochotona himalayana	Himalayan pika	China and Nepal
Ochotona hoffmanni	Hoffmann's pika	Mongolia, Eastern Russia
Ochotona huangensis	Tsing-Ling pika	China
Ochotona hyperborea	northern pika	North Asia
Ochotona iliensis	Ili pika	China
Ochotona koslowi	Kozlov's pika	China
Ochotona ladacensis	Ladak pika	China, India, Pakistan
Ochotona macrotis	large-eared pika	North, South, and Central Asia
Ochotona muliensis	Muli pika	China
Ochotona nigritia	black pika	China
Ochotona nubrica	Nubra pika	China, India, Nepal
Ochotona pallasi	Pallas's pika	Central Asia
Ochotona princeps	American pika	Western North America
Ochotona pusilla	steppe pika	Central Asia

Ochotona roylei	Royle's pika	Pakistan, India, Nepal, China
Ochotona rufescens	Afghan pika	Central Asia
Ochotona rutila	Turkestan red pika	Central Asia
Ochotona thibetana	Moupin pika	China, India, Burma
Ochotona thomasi	Thomas's pika	China
Ochotona turuchanensis	Turuchan pika	Eastern Russia

Family Leporidae
Rabbits

Brachylagus idahoensis	pygmy rabbit	Northwest United States
Bunolagus monticularis	riverine rabbit	South Africa
Caprolagus hispidus	bristly rabbit	India, Nepal, Bangladesh
Nesolagus netscheri	Sumatran striped rabbit	Sumatra (Indonesia)
Nesolagus timminsi	Annamite striped rabbit	Vietnam, Laos
Oryctolagus cuniculus	European rabbit	Iberian Peninsula (natural range)
Pentalagus furnessi	Amami rabbit	Japan
Poelagus marjorita	Bunyoro rabbit	Central Africa
Pronolagus crassicaudatus	Natal red rock rabbit	Southern Africa
Pronolagus randensis	Jameson's red rock rabbit	Southern Africa
Pronolagus rupestris	Smith's red rock rabbit	Southern and Eastern Africa
Pronolagus saundersiae	Hewitt's red rock rabbit	Southern Africa
Romerolagus diazi	volcano rabbit	Mexico
Sylvilagus aquaticus	swamp rabbit	South Central United States
Sylvilagus audubonii	desert cottontail	Western United States, Mexico
Sylvilagus bachmani	brush rabbit	Western United States, Mexico
Sylvilagus brasiliensis	tapeti	South America
Sylvilagus cognatus	Manzano Mountain cottontail	New Mexico
Sylvilagus cunicularius	Mexican cottontail	Mexico
Sylvilagus dicei	Dice's cottontail	Panama and Costa Rica
Sylvilagus floridanus	eastern cottontail	Southern Canada to Northern South America
Sylvilagus gabbi	Gabb's cottontail	Central America
Sylvilagus graysoni	Tres Marías cottontail	Tres Marías Islands (Mexico)
Sylvilagus insonus	Omilteme rabbit	Mexico

Sylvilagus mansuetus	San Jose brush rabbit	Mexico
Sylvilagus nuttallii	mountain cottontail	Western United States and Canada
Sylvilagus obscurus	Appalachian cottontail	Eastern United States
Sylvilagus palustris	marsh rabbit	Southeastern United States
Sylvilagus robustus	robust cottontail	Texas, New Mexico, Northern Mexico
Sylvilagus transitionalis	New England cottontail	Northeastern United States
Sylvilagus varynaensis	Venezuelan lowland rabbit	Venezuela

Hares

Lepus alleni	antelope jackrabbit	Arizona
Lepus americanus	snowshoe hare	Northern North America
Lepus arcticus	arctic hare	Alaska, Canada, Greenland
Lepus brachyurus	Japanese hare	Japan
Lepus californicus	black-tailed jackrabbit	Western North America
Lepus callotis	white-sided jackrabbit	Mexico, New Mexico
Lepus capensis	Cape hare	Africa, Asia Minor
Lepus castroviejoi	broom hare	Spain
Lepus comus	Yunnan hare	China
Lepus coreanus	Korean hare	Korea, China
Lepus corsicanus	Corsican hare	South Italy, Sicily (introduced to Corsica)
Lepus europaeus	European hare	Europe, Parts of Asia
Lepus fagani	Ethiopian hare	Ethiopia, Sudan, Kenya
Lepus flavigularis	Tehuantepec jackrabbit	Western Mexico
Lepus granatensis	Granada hare	Iberian Peninsula, Mallorca
Lepus habessinicus	Abyssinian hare	Horn of Africa
Lepus hainanus	Hainan hare	Hainan (China)
Lepus insularis	black jackrabbit	Espiritu Santo Island (Mexico)
Lepus microtis	African savanna hare	Africa
Lepus mandshuricus	Manchurian hare	Russian Far East, Northeast China, Northeast Korea

Appendix

Lepus melainus	Manchurian black hare	Russian Far East, Northeast China, Northeast Korea
Lepus nigricollis	Indian hare	Indian Subcontinent
Lepus oiostolus	woolly hare	India, Nepal, China
Lepus othus	Alaskan hare	Alaska, Eastern Russia
Lepus peguensis	Burmese hare	Southeast Asia
Lepus saxatilis	scrub hare	Southern Africa
Lepus sinensis	Chinese hare	Southeast China, Taiwan, Vietnam
Lepus starcki	Ethiopian highland hare	Central Ethiopia
Lepus tibetanus	desert hare	Central Asia
Lepus timidus	mountain hare	Northern Eurasia
Lepus tolai	Tolai hare	Central Asia
Lepus townsendii	white-tailed jackrabbit	Western North America
Lepus yarkandensis	Yarkand hare	China

Bibliography

The references below are a selected bibliography. A complete, unedited bibliography can be found at the Web page for this title at www.jhu.press.edu.

Alexander, R. M. 2003. *Principles of Animal Locomotion.* Princeton: Princeton University Press.

Alves, P. C., J. Melo-Ferreira, H. Freitas, and P. Bourset. 2008. The ubiquitous mountain hare mitochondria: Multiple introgressive hybridization in hares, genus *Lepus. Philosophical Transactions of the Royal Society Series B* 363:2831–2839.

Alzaga, V., J. Vicente, D. Villanua, P. Acevedo, F. Casas, and C. Gortazar. 2008. Body condition and parasite intensity correlates with escape capacity in Iberian hares (*Lepus granatensis*). *Behavioral Ecology and Sociobiology* 62:769–775.

Angelici, F. M., and L. Luiselli. 2007. Body size and altitude partitioning of the hares *Lepus europeaus* and *L. corsicanus* living in sympatry and allopatry in Italy. *Wildlife Biology* 13:251–257.

Angerbjorn, A., and J. E. C. Flux. 1995. *Lepus timidus. Mammalian Species* 495:1–11.

Asher, R. J., J. Meng, J. R. Wible, M. C. McKenna, G. W. Rougier, D. Dashzeveg, and M. J. Novacek. 2005. Stem Lagomorpha and the antiquity of Glires. *Science* 307:1091–1094.

Barrientos, R., and L. Bolonio. 2009. The presence of rabbits adjacent to roads increases polecat road mortality. *Biodiversity Conservation* 18:405–418.

Bell, D. J. 1985. The rabbits and hares: Order Lagomorpha. In *Social Odours in Mammals*, edited by R. E. Brown and D. W. Macdonald, 2:507–530. Oxford, United Kingdom: Clarendon Press.

Ben Slimen, H., F. Z. Suchentrunk, C. Stamatis, Z. Mamuris, H. Sert, P. C. Alves, U. Kryger, A. B. Shahin, and A. Ben Ammar Elgaaied. 2008. Population genetics of cape and brown hares (*Lepus capensis and L. europaeus*): A test of Petter's hypothesis of conspecificity. *Biochemical Systematics and Ecology* 36:22–39.

Bergstrom, D. M., A. Lucieer, K. Kiefer, J. Wasley, L. Belbin, T. K. Pedersen, and S. L. Chown. 2009. Indirect effects of invasive species removal devastate World Heritage Island. *Journal of Applied Ecology* 46:73–81.

Best, T. 1996. *Lepus californicus. Mammalian Species* 530:1–10.

Best, T. L., and T. H. Henry. 1993a. *Lepus alleni. Mammalian Species* 424:1–8.

Best, T. L., and T. H. Henry. 1993b. *Lepus callotis. Mammalian Species* 442:1–6.

Best, T. L., and T. H. Henry. 1994a. *Lepus arcticus. Mammalian Species* 457:1–9.

Best, T. L., and T. H. Henry. 1994b. *Lepus othus. Mammalian Species* 458:1–5.

Bósze, Z. S., and L. M. Houdebine. 2006. Application of rabbits in biomedical research: A review. *World Rabbit Science* 14:1–14.

Bover, P., J. Quintana, and J. A. Alcover. 2008. Three islands, three worlds: Paleogeography and evolution of the vertebrate fauna from the Balearic Islands. *Quaternary International* 182:135–144.

Branco, M., M. Monnerot, N. Ferrand, and A. R. Templeton. 2002. Postglacial dispersal of the European rabbit (*Oryctolagus cuniculus*) on the Iberian Peninsula reconstructed from nested clade and mismatch analyses of mitochondrial DNA genetic variation. *Evolution* 56:792–803.

Brewer, N. R. 2006. Biology of the rabbit. *Journal of the American Association for Laboratory Science* 45:8–24.

Burton, C. 2002. Microsatellite analysis of multiple paternity and male reproductive success in the promiscuous snowshoe hare. *Canadian Journal of Zoology* 80:1948–1956.

Caro, T. 2005. *Antipredator Defenses in Birds and Mammals.* Chicago: University of Chicago Press.

Cervantes, F. A. 1993. *Lepus flavigularis. Mammalian Species* 423:1–3.

Cervantes, F. A. 1997. *Sylvilagus graysoni. Mammalian Species* 559:1–3.

Cervantes, F. A., and C. Lorenzo. 1997. *Sylvilagus insonus. Mammalian Species* 568:1–4.

Cervantes, F. A., C. Lorenzo, and R. S. Hoffmann. 1990. *Romerolagus diazi. Mammalian Species* 360:1–7.

Cervantes, F. A., C. Lorenzo, J. Vargas, and T. Holmes. 1992. *Sylvilagus cunicularius. Mammalian Species* 412:1–4.

Chapman, J. A. 1974. *Sylvilagus bachmani. Mammalian Species* 34:1–4.

Chapman, J. A. 1975a. *Sylvilagus nutallii. Mammalian Species* 56:1–3.

Chapman, J. A. 1975b. *Sylvilagus transitionalis. Mammalian Species* 55:1–4.

Chapman, J. A., and G. A. Feldhammer. 1981. *Sylvilagus aquaticus. Mammalian Species* 151:1–4.

Chapman, J. A., and J. E. C. Flux, eds. 1990. *Rabbits, Hares and Pikas: Status Survey and Conservation Action Plan.* Gland, Switzerland: International Union for the Conservation of Nature.

Chapman, J. A., J. G. Hockman, and M. M. Ojeda. 1980. *Sylvilagus floridanus. Mammalian Species* 136:1–8.

Chapman, J. A., and G. R. Willner. 1978. *Sylvilagus audubonii. Mammalian Species* 106:1–4.

Chapman, J. A., and G. R. Willner. 1981. *Sylvilagus palustris. Mammalian Species* 153:1–3.

Cheeke, P. R. 1987. *Rabbit Feeding and Nutrition.* Orlando, FL: Academic Press.

Corbet, G. B. 1982. The occurrence and significance of the pectoral mane in rabbits and hares. *Journal of Zoology* 198:541–545.

Cowan, D. P. 1987. Group living in the European rabbit (*Oryctolagus cuniculus*): Mutual benefit or resource localization? *Journal of Animal Ecology* 56:779–795.

Dahl, F., and T. Willebrand. 2005. Natal dispersal, adult home ranges and site fidelity of mountain hares *Lepus timidus* in the boreal forest of Sweden. *Wildlife Biology* 11:309–317.

Davis, C. M., and V. L. Roth. 2008. The evolution of sexual size dimorphism in cottontail rabbits (*Sylvilagus*, Leporidae). *Biological Journal of the Linnean Society* 95:141–156.

Davis, S. E., and M. Demello. 2003. *Stories Rabbits Tell: A Natural and Cultural History of a Misunderstood Creature.* New York: Lantern Books.

Dearing, M. D. 1997. The manipulation of plant toxins by a food-hoarding herbivore, *Ochotona princeps. Ecology* 78:774–781.

Dearing, M. D., W. J. Foley, and S. McLean. 2005. The influence of plant secondary metabolites on the nutritional ecology of herbivorous terrestrial mammals. *Annual Review of Ecology and Systematics* 36:169–189.

Delibes-Mateos M., M. Delibes, P. Ferreras, and R. Villafuerte. 2008. Key role of European rabbits in the conservation of the Western Mediterranean basin hotspot. *Conservation Biology* 22:1106–1117.

DeStefano, S., S. L. Schmidt, and J. C. deVos Jr. 2000. Observations of predator activity at wildlife water developments in southern Arizona. *Journal of Range Management* 53:255–258.

Dobson, F. S., A. T. Smith, and X. G. Wang. 2000. The mating system and gene dynamics of plateau pikas. *Behavioural Processes* 51:101–110.

Durant, P., and M. A. Guevara. 2000. Reproduction and productivity in *Sylvilagus varynaensis*, a lowland Venezuelan rabbit. *Zoocriaderos* 3:1–10.

Durant, P., and M. A. Guevara. 2001. A new rabbit species (*Sylvilagus*, Mammalia: Leporidae) from the lowlands of Venezuela. *Revista de Biología Tropical* 49: 369–381.

Eisenberg, J. F. 1981. *The Mammalian Radiations*. Chicago: University of Chicago Press.

Elias, B. A., L. A. Shipley, R. D. Sayler, and R. S. Lamson. 2006. Mating and parental care in captive pygmy rabbits. *Journal of Mammalogy* 87:921–928.

Estes-Zumpf, W. A., and J. L. Rachlow. 2009. Natal dispersal by pygmy rabbits (*Brachylagus idahoensis*). *Journal of Mammalogy* 90:363–372.

Farías, V., T. K. Fuller, F. A. Cervantes, and C. Lorenzo. 2006. Home range and social behavior of the endangered Tehuantepec jackrabbit (*Lepus flavigularis*) in Oaxaca, Mexico. *Journal of Mammalogy* 87:748–756.

Fitzner, R. W., and R. H. Gray. 1991. The status, distribution and ecology of wildlife on the U.S. DOE Hanford Site: A historical overview of research activities. *Environmental Monitoring and Assessment* 18:173–202.

Flux, J. E. C., and P. L. Fullagar. 1992. World distribution of the rabbit *Oryctolagus cuniculus* on islands. *Mammal Review* 22:151–205.

Forman, R. T. T., and L. E. Alexander. 1998. Roads and their major ecological effects. *Annual Review of Ecology and Systematics* 29: 207–231+C2.

Franken, R. J., and D. S. Hik. 2004. Interannual variation in timing of parturition and growth of collared pikas (*Ochotona collaris*) in the southwest Yukon. *Integrative and Comparative Biology* 44:186–193.

Frey, J. K., R. D. Fisher, and L. A. Ruedas. 1997. Identification and restriction of the type locality of the Manzano Mountains cottontail, *Sylvilagus cognatus* Nelson, 1907 (Mammalia: Lagomorpha: Leporidae). *Proceedings of the Biological Society of Washington* 110:329–331.

Grayson, D. K. 2006. The Late Quaternary biogeographic histories of some Great Basin mammals (western USA). *Quaternary Science Reviews* 25:2964–2991.

Green, J. S., and J. T. Flinders. 1980. *Brachylagus idahoensis*. *Mammalian Species* 125:1–4.

Griffins, P. C., S. C. Griffins, C. Waroquiers, and L. S. Mills. 2005. Mortality by moonlight: Predation risk and the snowshoe hare. *Behavioral Ecology* 16:938–944.

Hackländer, K., W. Arnold, and T. Ruf. 2002. Postnatal development and thermoregulation in the precocial European hare (*Lepus europaeus*). *Journal of Comparative Physiology B*. 172:183–190.

Hackländer, K., F. Tataruch, and T. Ruf. 2002. The effect of dietary fat content on lactation energetics of the European hare (*Lepus europaeus*). *Physiological and Biochemical Zoology* 75:19–28.

Harcourt-Brown, F. 2002. *Textbook of Rabbit Medicine*. Oxford, UK: Butterworth-Heinemann.

Hoekstra, H. E. 2006. Genetics, development and evolution of adaptive pigmentation in vertebrates. *Heredity* 97:222–234.

Hughes, G. O., W. Thuiller, G. F. Midgley, and K. Collins. 2008. Environmental change hastens the demise of the critically endangered riverine rabbit (*Bunolagus monticularis*). *Biological Conservation* 141:23–34.

Janis, C. M., G. F. Gunnell, and M. D. Uhen, eds. 2007. *Evolution of Tertiary Mammals of North America*, vol. 2. Cambridge: Cambridge University Press.

Kamler, J. F. 2008. Ear flashing behaviour of cape hares (*Lepus capensis*) in South Africa. *African Journal of Ecology* 46: 443–444.

Kingdon, J. 1997. *The Kingdon Field Guide to African Mammals*. San Diego, CA: Academic Press.

Kraatz, B. P., D. Badamgarav, and F. Bibi. 2009. *Gomphos ellae*, a new mimotonid from the Middle Eocene of Mongolia and its implications for the origin of Lagomorpha. *Journal of Vertebrate Paleontology* 29:576–583.

Krebs, C. J., S. Boutin, and R. Boonstra, eds. 2001. *Ecosystem Dynamics of the Boreal Forest: The Kluane Project*. Oxford: Oxford University Press.

LaFever, D. H., R. R. Lopez, R. A. Feagin, and N. J. Silvy. 2007. Predicting the impacts of future sea-level rise on an endangered lagomorph. *Environmental Management* 40:430–437.

Lai, Y.-C., T. Shiroishi, K. Moriwaki, M. Motokawa, and H.-T. Yu. 2008. Variation of coat color in house mice throughout Asia. *Journal of Zoology* 274:270–276.

Larrucea, E. S., and P. F. Brussard. 2008. Shift in location of pygmy rabbits (*Brachylagus idahoensis*) habitat in response to changing environments. *Journal of Arid Environments* 72:1636–1643.

Laska, M. 2002. Gustatory responsiveness to food-associated saccharides in European rabbits, *Oryctolagus cuniculus*. *Physiology & Behavior* 76:335–341.

Laundré, J. W., J. M. M. Calderas, and L. Hernández. 2009. Foraging in the landscape of fear, the predator's dilemma: Where should I hunt? *The Open Ecology Journal* 2:1–6.

Lebas, F., P. Coudert, H. de Rochambeau, and R. G. Thébault. 1997. The Rabbit—Husbandry, Health and Production, rev. ed. FAO Animal Production and Health Series, no. 21. FAO, Rome, Italy. www.fao.org/docrep/t1690e/t1690e00.htm.

Lees, A. C., and D. J. Bell. 2008. A conservation paradox for the 21st century: The European wild rabbit *Oryctolagus cuniculus*, an invasive alien and an endangered native species. *Mammal Review* 38:304–320.

Liao, L., Z. Zhang, and N. Liu. 2007. Effects of altitudinal change on the auditory bulla in *Ochotona daurica* (Mammalia, Lagomorpha). *Journal of Zoological Systematics and Evolutionary Research* 45:151–154.

Lim, B. K. 1987. *Lepus townsendii*. *Mammalian Species* 288:1–6.

Litvaitis, J. A., and J. P. Tash. 2005. New England Cottontail, *Sylvilagus transitionalis*. In *New Hampshire Wildlife Action Plan*, A–303-A–312. New Hampshire Fish and Game Department, Concord, NH. http://extension.unh.edu/resources/files/Resource001071_Rep1313.pdf

Lockley, R. M. 1961. Social structure and stress in the rabbit warren. *Journal of Animal Ecology* 30:385–423.

Lombardi, L., N. Fernández, S. Moreno, and R. Villafuerte. 2003. Habitat-related differences in rabbit (*Oryctolagus cuniculus*) abundance, distribution, and activity. *Journal of Mammalogy* 84:26–36.

Lomolino, M. V. 2005. Body size evolution in insular vertebrates: Generality of the island rule. *Journal of Biogeography* 32:1683–1699.

MacDonald, S. O., and C. Jones. 1987. *Ochotona collaris*. *Mammalian Species* 281:1–4.

Marín, A. I, L. Hernández, and J. W. Laundré. 2003. Predation risk and food quantity in the selection of habitat by black-tailed jackrabbit (*Lepus californicus*): An optimal foraging approach. *Journal of Arid Environments* 55:101–110.

Marsden, H. M., and N. R. Holler. 1964. Social behavior in confined populations of the cottontail and the swamp rabbit. *Wildlife Monographs* 13:3–39

Matthee, C. A., B. J. Van Vuuren, D. Bell, and T. J. Robinson. 2004. A molecular supermatrix of the rabbits and hares (Leporidae) allows for the identification of five intercontinental exchanges during the Miocene. *Systematic Biology* 53:433–447.

McNab, B. K. 2002. *The Physiological Ecology of Vertebrates: A View from Energetics*. Cornell University Press, Ithaca, New York.

Meng, J. 2004. Chapter 7: Phylogeny and divergence of basal Glires. *Bulletin of the American Museum of Natural History* 285:93–109.

Monclús, R., F. Palomares, Z. Tablado, A. Martinez-Fontúrbel, and R. Palme. 2009. Testing the threat-sensitive predator avoidance hypothesis: Physiological responses and predator pressure in rabbits. *Physiological Ecology* 158:615–623.

Niu, Y., F. Wei, M. Li, X. Liu, and Z. Feng. 2004. Phylogeny of pikas (Lagomorpha, *Ochotona*) inferred from mitochondrial cytochrome *b* sequences. *Folia Zoologica* 53:141–155.

Olson, S. L. 1996. History and ornithological journals of the *Tanager* expedition of 1923 to the Northwestern Hawaiian Islands, Johnston and Wake islands. *Atoll Research Bulletin* 433:1–206.

Rachlow, J. L., D. M. Sanchez, and W. A. Estes-Zumpf. 2005. Natal burrows and nests of free-ranging pygmy rabbits (*Brachylagus idahoensis*). *Western North American Naturalist* 65:136–139.

Reichlin, T., E. Klansek, and K. Hackländer. 2006. Diet selection by hares (*Lepus europaeus*) in arable land and its implications for habitat management. *European Journal of Wildlife Research* 52:109–118.

Robinson, T. J., and C. A. Matthee. 2005. Phylogeny and evolutionary origins of the Leporidae: A review of cytogenics, molecular analyses and a supermatrix analysis. *Mammal Review* 35:231–247.

Rödel, H. G., A. Starkloff, A. Bautista, A.-C. Friedrich, and D. von Holst. 2008. Infanticide and maternal offspring defence in European rabbits under natural breeding conditions. *Ethology* 114:22–31.

Rödel, H. G., D. von Holst, and C. Kraus. 2009. Family legacies: Short- and long-term fitness consequences of early-life conditions in female European rabbits. *Journal of Animal Ecology* 78:789–797.

Rose, K. D., V. B. DeLeon, P. Missiaen, R. S. Rana, A. Sahni, L. Singh, and T. Smith. 2007. Early Eocene lagomorph (Mammalia) from Western India and the early diversification of Lagomorpha. *Proceedings of the Royal Society B* 275:1203–1208.

Ruedas, L. A. 1998. Systematics of *Sylvilagus* Gray, 1867 (Lagomorpha: Leporidae) from southwestern North America. *Journal of Mammalogy* 79:1355–1378.

Ruedas, L. A. and J. Salazar-Bravo. 2007. Morphological and chromosomal taxonomic assessment of *Sylvilagus brasiliensis gabbi* (Leporidae). *Mammalia* 71:63–69.

Schmidt-Nielsen, K. 1964. *Desert Animals: Physiological Problems of Heat and Water.* Oxford: Oxford University Press.

Seccombe-Hett, P., and R. Turkington. 2008. Summer diet selection of snowshoe hares: A test of nutritional hypotheses. *Oikos* 117:1874–1884.

Shipley, L. A., T. B. Davila, N. J. Thines, and B. A. Elias. 2006. Nutritional requirements and diet choices of the pygmy rabbit (*Brachylagus idahoensis*): A sagebrush specialist. *Journal of Chemical Ecology* 32:2455–2474.

Smith, A. T. 2003. Lagomorpha. In *Grzimek's Animal Life Encyclopedia*, vol. 16 Mammals, part 5, 479–516. Farmington Hills, MI: Thomson Gale.

Smith, A. T., and J. M. Foggin. 1999. The plateau pika (*Ochotona curzoniae*) is a keystone species for biodiversity on the Tibetan plateau. *Animal Conservation* 2:235–240.

Smith, A. T., and M. L. Weston. 1990. *Ochotona princeps. Mammalian Species* 352:1–8.

Smith, A. T. and Y. Xie. 2008. *A Guide to the Mammals of China*. Princeton: Princeton University Press.

Somers, N., B. D'Haese, B. Bossuyt, L. Lens, and M. Hoffmann. 2008. Food quality affects diet preference of rabbits: Experimental evidence. *Belgium Journal of Zoology* 138:170–176.

Stiner, M. C., N. D. Munro, and T. A. Surovell. 2000. The tortoise and the hare: Small-game use, the broad-spectrum revolution, and paleolithic demography. *Current Anthropology* 41:39–73.

Stockely, P. 2003. Female multiple mating behaviour, early reproductive failure and litter size variation in mammals. *Proceedings of the Royal Society of London B* 270:271–278.

Stoner, C. J, O. R. P. Bininda-Emonds, and T. Caro. 2003. The adaptive significance of coloration in lagomorphs. *Biological Journal of the Linnean Society* 79:309–328.

Stott, P. 2008. Comparisons of digestive function between the European hare (*Lepus europaeus*) and the European rabbit (*Oryctolagus cuniculus*): Mastication, gut passable, and digestibility. *Mammalian Biology* 73:276–286.

Stott, P., N. Jennings, and S. Harris. 2010. Is the large size of the pinna of the ear of the European hare (*Lepus europaeus*) due to its role in thermoregulation or in anterior capital shock absorption? *Journal of Morphology.* Published online in Wiley InterScience (www.interscience.wiley.com). DOI: 10.1002/jmor.10825

Surridge, A. K., K. M. Ibrahim, D. J. Bell, N. J. Webb, C. Rico, and G. W. Hewitt.

2002. Fine-scale genetic structuring in a natural population of European wild rabbits (*Oryctolagus cuniculus*). *Molecular Ecology* 8:299–307.

Swihart, R. K. 1984. Body size, breeding season length, and life history tactics of lagomorphs. *Oikos* 43: 282–290.

Tablado, Z., E. Revilla, and F. Palomares. 2009. Breeding like rabbits: Global patterns of variability and determinants of European wild rabbit reproduction. *Ecography* 32:310–320.

Thomas, H. H., and T. L. Best. 1994a. *Lepus insularis. Mammalian Species* 465:1–3.

Thomas, H. H., and T. L. Best. 1994b. *Sylvilagus mansuetus. Mammalian Species* 464:1–2.

Thompson, H. V., and C. M. King. 1994. *The European Rabbit: The History and Biology of a Successful Colonizer.* Oxford: Oxford University Press.

Trefry, S. A., and D. S. Hik. 2009. Variation in pika (*Ochotona collaris, O. princeps*) vocalizations within and between populations. *Ecography* doi:10.1111/j.1600 –0587.2009.05589.x.

van der Loo, W., J. Abrantes, and P. J. Esteves. 2009. Sharing of endogenous lentiviral gene fragments among leporid lineages separated for more than 12 million years. *Journal of Virology* 83:2386–2388.

Virgós, E., S. Cabezas-Díaz, and J. A. Blanco-Aguiar. 2006. Evolution of life history traits in Leporidae: A test of nest predation and seasonality hypotheses. *Biological Journal of the Linnean Society* 88:603–610.

von Holst, D., H. Hutzelmeyer, P. Kaetzke, M. Khaschei, and R. Schonheiter. 1999. Social rank, stress, fitness, and life expectancy in wild rabbits. *Naturwissenschaften* 86:388–393.

Ward, D. 2005. Reversing rabbit decline: One of the biggest conservation challenges for nature conservation in Spain and Portugal. www.deshok.com/soslynxfile/ info_file/reversing.pdf.

Wible, J. R., G. W. Rougier, M. J. Novacek, and R. J. Asher. 2007. Cretaceous eutherians and Laurasian origin for placental mammals near the K-T boundary. *Nature* 447:1003–1006.

Wilson, D. E., and D. M. Reeder, eds. 2005. *Mammal Species of the World: A Taxonomic and Geographic Reference*, vol. 1, 3rd ed. Baltimore: Johns Hopkins University Press.

Yamada, F., and F. A. Cervantes. 2005. *Pentalagus furnessi. Mammalian Species* 782:1–5.

Yang, J., Z. L. Wang, X. Q. Zhao, D. P. Wang, D. L. Qi, B. H. Xu, Y. H. Ren, and H. F. Tian. 2008. Natural selection and adaptive evolution of leptin in the *Ochotona* family driven by the cold environmental stress. *PLoS ONE* 3: e1472 .doi:10.1371/journal.pone.0001472.

Yin, B., S. Yang, W. Wei, and Y. Zhang. 2009. Male reproductive success in plateau pikas (*Ochotona curzoniae*): A microsatellite analysis. *Mammalian Biology* 74:344–350.

Yu, N., C. Zheng, Y.-P. Zhang, and W.-H. Li. 2000. Molecular systematics of pikas (genus *Ochotona*) inferred from mitochondrial DNA sequences. *Molecular Phylogenetics and Evolution* 16: 85–95.

Index

adults, 54, 61, 65, 70, 71, 86, 136, 188; and
   growth, 126–27; and life span, 129, 130; and
   mortality, 69, 129–30; and play, 73, 74
Aesop, 200
Africa, 13, 14, 17, 80, 91, 94, 100, 211
aggression, 52, 67, 69, 72, 125, 148, 154, 171;
   and fighting, 70–71; and mating, 65, 111, 112
agriculture, 16, 108, 166, 167, 183, 197; in
   Australia, 163, 165; and habitat loss, 172,
   179, 194; and herbicides, 183
Alaska, 89, 94
albinism, 56, 61–63
Alilepus, 26
Allen's rule, 42–43
Alps, 94, 179
altitude/elevation, 91, 92, 178; and breeding
   season, 117, 118; and coloration, 52, 56, 58;
   and litter size, 119; and pikas, 15; and size, 30
American Rabbit Breeders Association, 145,
   213
anal glands, 77–78, 79
Animal Welfare Act, 152, 154
Annamite Mountains, 9, 10, 11
Asia, 1, 13–14, 19, 20, 24, 80, 90–91
Australia, xiii, 14, 94, 120, 162, 186, 193; and
   disease, 103, 104; management of rabbits in,
   163–65, 209–10; predators of, 16, 43; rabbit-
   proof fence, 163-64, 165

Balearic Islands, 159, 160
bears, 15, 80, 108, 166
birds, 161–62; nesting by, 108; predation by, 15,
   16, 79, 81–82, 87, 108, 140, 193; and water
   sites, 140. *See also* buzzards; eagles; owls,
   ravens; vultures
bobcats, 15, 16, 80, 83, 137, 193
body size, 4, 6, 29–31, 32, 40, 42, 72, 85; and
   fossils, 26–28; and life span, 128; and litter
   size, 118, 119; and thermoregulation, 96,
   97, 98
body temperature, 32, 47, 96, 98, 99, 114, 127;
   and coloration, 49, 63; and desert conditions,
   95–96, 97; and ear size, 42–43, 44, 47, 96;
   and litter size, 114, 119, 120; and nursing,
   123; and sociality, 68; and young, 116, 123,
   124-25
body weight, 3, 73, 101, 115, 123, 124, 140;
   and brown fat, 99, 152; of different species,
   12, 26, 29–30, 97; and growth, 126–28; and
   herbicides, 182; and litter size, 120; of males
   and females, 31
boxing, 70, 111

Brachylagus, 12
brain, 35, 36, 39, 59, 62, 72–73, 148, 154
Br'er Rabbit, 201
Bugs Bunny, 201
Bunnicula (Howe), 207
Bunolagus, 13
burrows, 68, 80, 95, 99, 137, 146, 147, 164;
   birth and rearing in, 115, 119, 124, 129;
   danger from, 160, 161; and European rabbits,
   13, 14, 44, 64, 79, 86, 210; formation and use
   of, 6, 40, 84, 86, 87; and running, 44, 46; as
   shelter for other species, 108, 109, 110. *See
   also* forms; nests
buzzards, 82, 108

California, 94, 193
Canada, 13, 94, 184
canids, 79, 80, 81. *See also* coyotes; foxes; wolves
Caprolagus, 13
carnivores, 3, 19, 21, 87, 131, 169, 171, 193;
   coloration of, 49; and encephalization
   quotient, 72; food for, 18, 108, 109;
   metabolism of, 32; predation by, 79–83, 87;
   and sleep, 40
Carter, Jimmy, attacked by rabbit, 170, 171
Caspian Sea, 94
Caswell Memorial State Park, 175
cats, 19, 21, 63, 71, 80, 152, 169; and Australia,
   16, 163, 164; avoidance of, 87; breeding
   of, 145; domestic, 148; eyesight of, 41;
   feral domestic, 193; food for, 15, 79; and
   Macquarie Island, 162; and parasites, 101;
   and perception, 36, 41; and *Toxoplasma gondii*,
   101; and Tularemia, 170. *See also* bobcats;
   lynx
Chile, 14, 80, 104
China, 9, 103, 104, 146, 166, 178, 194; culture
   of, 189, 190, 197, 198, 199, 200; and rabbits
   as food, 185, 188; species in, 15, 20, 25, 90,
   92, 94
clade, crown, 19, 22
climate, 15, 18, 31, 32, 164; and Allen's rule,
   42–43; and breeding season, 117, 118, 162;
   effects of change in, 172, 176–81, 183, 212;
   and global warming, 176–77, 178, 194;
   historical change in, 14, 18, 27
communication, 6, 35, 44, 49, 74–77, 77–79
conservation, 18–19, 93, 165, 173–76, 186–87,
   193–95, 209
cottontail, desert, 21, 32, 43, 48, 61, 89,
   90, 140; and desert habitat, 95, 96; and
   thermoneutrality, 98

cottontail, eastern, 9, 20, 21, 65, 89, 173, 176, 210; and affection, 67; and breeding season, 117; and chin mark, 78; and climate change, 180; coloration of, 51; diet of, 134; and disease, 102; distress call, 74; distribution of, 93, 94; and food, 85; fur of, 59, 61; and gestation, 114; growth rate of, 126; and habitat, 90; and home range, 86; latrines of, 78; and life span, 129; and litter size, 119, 120; nests of, 115, 116; and parasites, 100; polyandry among, 121; swimming by, 48; and vocalization, 74

cottontail, New England, 9, 43, 59, 83, 114, 119, 180, 210; as endangered, 173, 174, 176, 195; hunting of, 186–87; and roads, 192; running by, 47; and shelter, 84–85

cottontails, 48, 52, 85, 89, 105, 124, 210; and breeding season, 117–18; and desert habitat, 95; and disease, 102; as endangered, 173; as food, 171, 184; and foot thumping, 77; fur of, 61; and gestation, 114; growth rate of, 126, 127; hunting of, 186, 187; and litter size, 119; nests of, 115, 116; North American, 6, 211; origin of, 24–25; and parasites, 100; as prey, 15, 17, 80, 81; running by, 44, 46–47; and shelter, 84; size of, 30, 31; and social behavior, 64–65, 67; South American, 102, 211; species of, 11, 20–21, 22, 23; and tail and ear size, 43; weaning of, 154–55. See also cottontails, types of

cottontails, types of: Appalachian, 9, 21, 89; Gabb's, 9; Manzano Mountain, 9; marsh, 81; Mexican, 30, 90, 114; mountain, 21, 43, 101, 114, 115, 140, 173, 181–82; Omilteme, 52, 55, 90, 93, 174, 211; robust, 9, 93, 174; swamp, 81; Tres Marías, 93, 139, 174. See also cottontail, desert; cottontail, eastern; cottontail, New England

coyotes, 80, 106, 137, 140

Darwin, Charles, 185, 215
de la Mare, Walter, 201
Denmark, 105
deserts, 1, 7, 11, 13, 32, 50, 89, 90; survival in, 43, 89, 95–98, 117; and water intake, 139, 140
digestion, 4, 32, 33–35, 47, 100, 101, 132–36, 138
diseases, 100–104, 149, 162, 165, 170, 172, 177; and experimental research, 150, 151, 152–53, 154, 208, 209
dispersion, 31, 48, 70, 74, 126, 192; and evolution, 11–12, 25; and sociality, 65–66, 67
distribution, 10, 11, 12, 88–91, 92–95
domestication, 146–49
dominance, 69, 70, 74, 78, 111, 112–13, 120, 121

eagles, 15, 16, 79, 81, 82, 87, 109, 164
ears, 70, 127, 213; and coloration, 53–54, 61; and heat, 47, 96, 97; size of, 4, 6, 42, 43, 44, 45, 127

Easter, 197
ecology, xiii, 18, 105, 194
elevation/altitude. See altitude/elevation
endangered species, 93–94, 172–76
Endangered Species Act (ESA), 173
Endangered Wildlife Trust, Riverine Rabbit Working Group, 175
energetics, and brain size, 73
environment, 73, 104–10, 172. See also climate
escape, 48, 74, 102, 114, 130, 181, 190–91; and roads, 190–91; strategies for, 6, 14, 32, 46; and trickster figure, 201
Espiritu Santo Island, 50, 51, 92
estrus, 70, 74, 78; origin of word, 197; and reproduction, 111, 112, 118, 120, 121, 123
Eurasia, 14, 25, 94, 170
Europe, 10, 11, 14, 15, 19, 80, 91, 94; conservation in, 174; diseases in, 103, 104; ecosystem of, 110; rabbits as food in, 16–17, 183, 185–86, 210; rabbit domestication in, 147–48
European Association of Rabbit Breeders, 145
evolution, 10, 11, 12, 19–28
experimentation, 150–53, 154, 208, 209
eyes, 6, 39, 128, 148, 154; of albinos, 61, 62–63; color of, 54, 61; and Draize test, 151–52; and fossil rabbits, 26, 27; glands around, 77, 78; and running, 44; structure of, 41, 42; of young, 114, 126, 128

feces, 77–78, 123, 137, 140, 144; and digestion, 4, 33, 34–35, 36; and latrines, 77–78, 109–10; and soil fertility, 108, 109–10
feet, 3–4, 24, 25, 26, 70, 189; drumming or thumping of, 76–77; growth rate of, 126, 127; and running, 31, 44, 45–46, 47
females, 78, 86, 120; and care of young, 65–67, 76, 79, 122–26, 154–55; and chewing, 136; dispersal of, 65, 66, 67; and fighting, 70–71; and infanticide, 71; and lactation, 79, 114, 117; and life span, 130; and mating, 65, 120–22; and nests, 115; and nursing, 3, 122–24, 125, 126; and ovulation, 112, 113, 122, 149; and reproduction, 111–14; reproductive anatomy of, 4–5; and scent glands, 77, 78, 79; size of, 31; and territoriality, 69; two uteri of, 5, 122; and vocalization, 74–75, 76
ferrets, 15, 80, 108, 146
fighting, 69, 70–71, 74, 112, 123
fleas, 100, 101, 102, 109
folklore, xiii, 196, 198, 201
food, 4, 33, 70, 84, 88, 106, 126, 131–35; and breeding period, 117; chewing of, 135–36; and desert, 95, 97–98; and home range, 86, 87; kinds of, 131–35; and litter size, 119; location of, 137–38; and metabolic rate, 32; and migration, 87, 88; and plant species, 85; rabbits as, 147, 148, 171, 172, 183–87, 188, 209; selection of, 133, 134, 138, 141, 143; and social behavior, 68–69; storage of, 132,

141–43; toxins in, 132–35, 138, 139; and vocalization, 75; and water intake, 139

forms, 6, 95, 96, 99, 116. *See also* burrows; nests

fossils, 19, 20, 22, 23–28, 208

foxes, 15, 16, 108, 140, 148, 193; arctic, 56, 80; corsac, 80; gray, 80; kit, 80; red, 58, 80, 81; swift, 80; Tibetan sand, 80

France, 14, 103, 149, 185

fur, 32, 61, 98, 99; color of, 49–54, 54–56, 56–60, 180–81; of domestic rabbits, 148; human use of, 147, 187–88, 209; and mammals, 3; molting of, 59; and nests, 115, 116; and thermoregulation, 97; and types of hairs, 49, 57

genetics, 9, 10, 12; and albinism, 61–62; and coloration, 54–56, 63; and disease, 102–3; of domestic rabbits, 149; and evolution, 19; experimental research on, 152; and rabbit genome, 209; and seasonal coloration changes, 58; and social behavior, 66; and vocalization, 75

Germany, 210

Glires, 2, 20, 22, 23, 24

Gloger's rule, 51–52

Gomphos elkema, 24

Gomphos ellae, 24

grazing, 68, 105, 106, 109, 110, 141, 159–67

Great Basin, 12, 178, 179

Great Britain, 85, 94, 103, 105, 160, 186, 210

Greeks, ancient, 186, 197, 199, 200

Gymnesicolagus, 27–28

habitat, 83; and Australia, 163, 164; and Baluchistan, 166; and brain size, 73; and climate change, 178–79, 180; and coloration, 50–52, 57, 58, 60; degradation and loss of, 15, 92, 93, 104, 109, 172, 173, 177, 191–92; desert, 95–98; and different groups, 7, 13; and distribution, 92–94; and ear size, 43–44; and European rabbits, 210; and evolution, 11; and foraging risk, 137; grass-dominated, 14, 25, 85; and grazing, 110; and home range, 86; and keystone species, 105; and metabolic rate, 32; and Miami Airport, 167; and migration, 87, 88; and nesting, 115, 116; protection of, 194–95; and shelter, 84, 86; and species diversity, 89–91; winter, 98-99

Hadrian's Wall, 160

Hanford Reach National Monument, Saddle Mountain National Wildlife Refuge, 182

hare, arctic, 22, 43, 48, 73, 74, 124, 129; and affection, 67; and breeding season, 118; in Canada, 89; coloration of, 57, 58; diet of, 131; fur of, 61, 98; and gestation, 114; and mating, 65, 112; metabolic rate of, 32, 98; and migration, 88; and nest, 116; predation on, 80, 81; and shelter, 84; speed of, 46; and water intake, 140; and winter, 98–99

hare, Cape, 43, 44, 80, 81, 91, 125, 211; and breeding season, 118; coloration of, 54; and desert habitat, 95, 96, 97, 98; and gestation, 115; and parasites, 101

hare, European, 11, 14, 22, 91, 102, 136, 210; as beneficial, 105; breeding of, 149; and Cape hare, 94, 97; and climate change, 179–80; diet of, 85, 131, 132, 134; dispersal of, 67; and distribution, 93; as endangered, 174, 183, 192; and gestation, 115; hunting of, 187; and litter size, 120; mating by, 65; and metabolism, 96; negative effects of, 104; and parasites, 100; and paternity, 120; predation on, 81; and shelter, 84; speed of, 46; and superfetation, 122; swimming by, 48; and thermoregulation, 116; and vocalization, 74; young of, 124, 125

hare, mountain, 21–22, 91, 115, 210; and breeding season, 117; in China, 90; and climate change, 179–80; diet of, 134; dispersal of, 67; distribution of, 94; as endangered, 173; growth rate of, 127; and home range, 86; and life span, 129, 130; and migration, 88; and parasites, 101; predation on, 80, 81; seasonal coloration changes of, 57, 58; and shelter, 84; and toxins, 182

hare, snowshoe, 15, 22, 23, 43, 89; and breeding season, 117; and Canada lynx, 80, 105–7; and climate change, 180–81; coloration of, 51, 57, 58, 59; diet of, 132, 133, 134; dispersal of, 67; distribution of, 94; and fighting, 70; and food selection, 138; and gestation, 114; and glyphosate, 182; hunting of, 186; and litter size, 119, 120; metabolic rate of, 32; and migration, 88; and parasites, 101; and paternity, 120; predation on, 16, 80, 81, 87; research on, 210, 211; and roads, 191; size of, 30, 31; swimming by, 48; and thermoneutrality, 96; and vocalization, 74; and water intake, 140; in winter, 98, 99; young of, 124

hares, xiii; distinguishing characteristics of, 3, 4–5, 6–7; evolution of, 21–25, 26; as lagomorphs, 1; species of, 8, 10, 14. *See also* hares, types of

hares, types of: Abyssinian, 91; African, 102; African savanna, 91; Alaskan, 30, 31, 57, 80, 81, 89, 116, 127; Belgian, 147, 151; broom, 11, 22, 91, 93–94, 174, 191; Burmese, 91; Corsican, 11, 91, 174, 187; desert, 90; Ethiopian, 91; Granada, 22, 91, 153; Hainan, 30, 92, 174, 184; hispid, 5; Iberian, 102; Indian, 50, 80, 91; Irish, 57, 173, 174, 210, 214; Japanese, 51, 57, 91; Manchurian, 52; Patagonian (mara), 6; scrub, 81, 91, 100, 193, 212; Tolai, 80, 90, 94, 101, 115; woolly, 59, 60, 220; Yarkand, 30, 31. *See also* hare, arctic; hare, Cape; hare, European; hare, mountain; hare, snowshoe

hay piles, 33, 75, 77, 126, 132, 133, 137, 141–43, 166

Index   231

hearing, 3, 26, 43–44, 54, 148
herbicides/pesticides, 169, 181–83, 194
herbivores, 33, 35, 37, 40, 105, 110, 132, 134
hibernation, 99–100
home range, 65, 77, 84, 85–87, 111, 117, 137
House Rabbit Society, 155
hunting, 18, 146, 148, 171, 173, 183–87,
    192, 209

Iberia, xiv, 15, 103, 109, 146, 186, 209
India, 25, 80, 189–90, 194, 200, 218
inguinal glands, 78–79
intelligence, 71–73
International Union for the Conservation of
    Nature (IUCN) Red List of Threatened
    Species, 172–73, 174, 209, 211
Islam, 184, 200
Italy, 91, 146, 185, 187

jackalopes, 200
jackrabbit, antelope, 43, 46; coloration of, 53;
    and desert habitat, 95–96; fur of, 61; and
    home range, 87; and parasites, 101; speed of,
    46; and vocalization, 74; and water intake,
    139, 140
jackrabbit, black-tailed, 5, 35, 43, 89, 210; and
    breeding season, 117; coloration of, 51, 53,
    54; and desert habitat, 95; distribution of, 94;
    ears of, 96; and food, 137, 138; fur of, 59, 61;
    growth rate of, 127; and habitat, 89; and life
    span, 129; mating by, 65; and migration, 88;
    and parasites, 100, 101; as pest, 166–67; and
    radioactive waste, 181–82; and reproduction,
    114; and roads, 191; speed of, 46; swimming
    by, 48; and thermoneutrality, 96; and vocal-
    ization, 74; and water intake, 140; in winter,
    98, 99
jackrabbits, 5; as food, 171, 184; and food selec-
    tion, 138; and gestation, 114–15; hunting of,
    186; lungs of, 46; nests of, 116; predation on,
    17, 80, 81; and reproduction, 111–12; and
    thermoneutrality, 98; in winter, 98
jackrabbits, types of: black, 50, 51, 92; Tehu-
    antepec, 14, 65, 89, 93, 174; white-sided, 43,
    53, 65, 89, 173, 174; white-tailed, 22, 42,
    48, 57, 58, 65, 89, 95, 98. See also jackrabbit,
    antelope; jackrabbit, black-tailed; jackrabbit,
    white-sided; jackrabbits, white-tailed
Japan, 13, 93, 94, 174, 193
Judaism, 184, 197, 200

kangaroo, 4, 44, 46, 164, 165

Lagomorph Specialist Group of the Interna-
    tional Union for the Conservation of Nature,
    208–9, 215
lagomorphs: characteristics of, 1–7; classifica-
    tion of, 5–15, 22; evolution of, 19–28
Lawrence, D. H., 204

Laysan Island, 160–62
Leporidae, 6–7, 20, 22, 24
Lepus, 5, 10, 14
Lepus (constellation), 199
life span, 128–30; and dominance, 69; and litter
    size, 118. *See also* mortality
Linnaeus, Carl, 1
literature, xiii, 1, 200–204
livestock, 109, 164, 166
Lockley, Ronald, 202
longevity. *See* life span
lynx: Canada, 15, 80, 87, 105–6, 181; Iberian,
    xiii, xiv, 15, 80, 109, 110, 209; and snowshoe
    hares, 210

Macquarie Island, 162–63
"mad as a March hare," 111
males: and care of young, 122; and chewing,
    136; and chin mark, 78; dispersal of, 65–66,
    67; and fighting, 70; and home range, 86; and
    latrines, 78; and life span, 130; and mating,
    65, 112, 120–22; and parent-offspring rela-
    tionships, 65; and reproduction, 111–13, 114;
    reproductive anatomy of, 4–5, 112, 113; and
    roads, 192; and scent marking, 77, 78; size
    of, 31; and territoriality, 69; and vocalization,
    74–75, 76
mammals: Afrotheria, 19, 20, 21; and brain
    size, 73; class of, 3–4; coloration of, 51; and
    crown clades, 19; and dinosaurs, 22; dispersal
    of, 65; and encephalization quotient, 72;
    Euarchontoglires, 20; evolution of, 19; and
    lagomorphs, 2–3; Laurasiatheria, 19, 20,
    21; and life span, 128; and litter size, 119;
    and metabolic rate, 32; placental, 4, 19; and
    roads, 190, 191; size of, 29; and sleep, 39,
    40; swimming by, 48; traits of domestic,
    148; typical color pattern of, 49; visual sys-
    tem of, 62; and whiskers, 44; Xenarthra, 19,
    20, 21
mating and reproduction, 65, 111–14, 117–22;
    acrobatic leaps in, 113; and anal glands, 78;
    and birth, 112, 115–17, 122; and body size,
    31; and brain size, 72; chases in, 111, 112,
    113; and chin mark, 78; and climate, 117; and
    coloration, 52; and copulation, 79, 111, 112,
    113, 120–21, 122; and dominance, 69, 113;
    and estrogen production, 113; and fertility,
    xiii, 202; and fighting, 70; and fitness, 112;
    gestation, 114–15, 151; and home range, 86;
    and latrines, 78; length of, 117–18, 119, 120;
    and litter size and numbers, 114, 118–20,
    127; and mammals, 3; and myth, 196, 197,
    198, 199; and parent-offspring relationships,
    65–67; and philopatry, 66, 67; and photope-
    riod, 117; and play, 74; and pregnancy, 124,
    125; promiscuous, 86, 120–21; and scent,
    77, 78–79; and seasonal breeding, 113–14;
    and species, 6, 8; and superfetation, 122; and

territoriality, 69; time of, 117–18; and urine spray, 79; and vocalization, 74, 75, 76

medicine, folk and traditional, 189–90

Mediterranean region, 17, 26, 81, 96, 109, 110

metabolism, 3, 31–32, 96, 98, 99

Mexico, 12, 13, 89–90, 92, 93, 94, 199; Maya, 197

mice, 1, 38, 51, 52, 152, 163

migration, 87–88

Mimotonidae, 24

Monty Python and the Holy Grail (film), 206-7

moon, myths about rabbits and, xiii, 196, 197, 198, 199

mortality, 70, 129, 130, 140, 190–92. *See also* life span

mythology, xiii, 1, 196–200

Native Americans, 17, 184, 188, 197, 199, 201

Nesolagus, 14

nest, 64, 69, 74, 114, 115–16, 117, 122, 124. *See also* burrows; nests

New Zealand, 14, 85, 94, 100, 162, 186, 193

Night of the Lepus (film), 205-6

North America, 14, 15, 19, 24, 25, 80, 94, 170, 186; boreal forest of, 17, 51, 88, 94, 105–7, 181, 184

Ochotona, 14, 20, 22

Ochotonidae, 6, 20, 22, 24

Oryctolagus, 14

Osiris, 196-97

Ostra (Eostre), 197

owls, 51, 79, 81, 82, 106, 107

pain, 152, 154

parasites, 100–102, 109, 112; ectoparasite, 109; endoparasite, 100. *See also* diseases

Pentalagus, 13

pests, rabbits as, 159–68

Peter Rabbit, 169, 201

pets, rabbit as, 144–46

photoperiod: and breeding season, 113, 117, 148; and coloration, 59, 149

pika, American, 65, 70, 85, 121, 126; and breeding season, 118; and climate change, 178, 179; dispersal of, 67; as endangered, 173; and food, 137, 138; fur of, 59, 61; and hay piles, 137, 141, 142, 143; and life span, 130; and litter size, 118–19; and scent marking, 77; and social behavior, 64; and vocalization, 75, 76; and winter habitat, 99

pikas: Asian, 76, 77, 101, 178; burrowing, 67, 84, 118, 125, 126, 130, 141; classification of, 2–3; distinguishing characteristics of, 1, 4, 5–7; evolution of, 20, 27–28; species of, 8, 9, 14–15, 20, 22; talus-living, 66, 84, 85, 86, 115, 118, 126, 130, 141–42

pikas, types of: Afghan, 76, 77, 81, 82, 141, 165–66; alpine, 29–30, 59, 77, 82, 86,

119–20, 121, 125; black, 9; Chinese red, 30; collared, 59, 76, 118, 119, 126, 131–32, 178, 210–11; Daurian, 43, 59, 82, 108; Gansu, 29, 30, 77; Hoffmann's, 174; Ili, 9, 174, 178; Kozlov's, 174; large-eared, 141; northern, 20, 30, 76, 77, 82, 95, 121; Pallas's, 20, 76, 77; plateau, 64, 65, 66, 107, 119, 121, 125–26, 130; Royle's, 82; Sardinian, 6, 15; silver, 92, 172, 174; steppe, 20; Thomas's, 29; Tsing-Ling, 29; Turkestan red, 76, 82. *See also* pika, American

pine-juniper woodland, 178, 179

play, 73–74

Playboy Bunny, 198, 202

Poelagus, 13

poison, 108, 109, 132–35, 166, 183

polecats, 80, 108, 191

pollution, 177, 181–83, 194

popular culture, 1, 6, 111, 131, 197, 198, 201–2, 204–7

Portugal, 103, 104, 109, 174, 186, 209

predation, 48, 71, 105, 108, 155, 179, 193; in Australia, 164; and care for young, 122, 125, 126; carnivores involved in, 79–83; and coloration, 49, 50, 51, 53–54, 57–58; and disease, 102; and ear size, 43; and eyesight, 41; and foot thumping, 77; and foraging risk, 137; and gestation, 114; and home range size, 85; and intelligence, 72; and life span, 130; and mating, 112, 121; and migration, 88; and pets, 144; and sheltering, 84, 87; and snow-shoe hare–lynx relationship, 106; and social behavior, 68–69; and vocalization, 75; and water intake, 140. *See also individual predators and prey*

primates, 3, 20, 37, 72

Prolagidae, 6

Pronolagus, 13

rabbit, Amami, 13, 22, 23, 48, 91, 100; coloration of, 51, 52; and distribution, 93; as endangered, 174; fur of, 54, 61; predation on, 81, 193; protection of, 174; and vocalization, 75

rabbit, domestic, 14, 145, 146–49, 215; and albinism, 61–62, 63; breeds of, 30, 145, 147–48; and chewing, 135; and communication, 79; consumption of, 171, 185–86, 188; and disease, 100, 101, 102, 103, 104; and experimental research, 150, 151, 154, 208, 209; fur of, 56, 59, 188, 189; growth of, 127, 128; and life span, 129; marking of, 213; and medicine, 190; and mother-young relationships, 122; and olfaction, 36; as pets, 144–46; and plant toxins, 133, 138; and play, 73; release in Australia, 163; release of, 17, 160; and Scott's tree kangaroo, 165; size of, 30; sleep patterns of, 39–40; teeth of, 38, 39; and water, 138, 139

rabbit, European, xiv, 6, 11, 13, 14, 17, 22, 23, 91; in Africa, 91; as beneficial, 105; as best known species, 209; and brain size, 73; and breeding season, 117, 118; and chewing, 136; and chin mark, 78; coloration of, 49, 52; and disease, 102, 104; and dispersal, 66; distribution of, 12, 94; domestic, 36, 38, 40–41, 59, 144–46, 209; and ear size, 43; and food, 37, 137, 138; and foot thumping, 76–77; and gestation, 114; growth rate of, 127; and home range, 86, 87; of Iberia, 109; and Iberian lynx, xiii, 80; and infanticide, 71; and intelligence, 72; and latrines, 78; and life span, 129; and litter size, 120; lungs of, 46; mating behavior, 113; molting of, 59; negative effects of, 104; nesting by, 64, 115, 117; and ovulation, 112; and parasites, 100, 101; and parent-offspring relationships, 65; as pests, 160; and plant species for food, 85; and polyandry, 121; predation on, 80, 81; as prey, 16, 193; research on, 209–10, 211; and roads, 191; running by, 44; and scent glands, 79; and shelter, 84; size of, 30; and sociality, 7, 64, 69; study of, 208; swimming by, 48; and urine spray, 79; and vocalization, 75; and water intake, 139, 140; and young, 124
rabbit, pygmy, 12, 22, 23, 28, 36, 45, 129; and birth, 115; and climate change, 178–79, 180; diet of, 134; dispersal of, 67; as endangered, 173, 174; fur of, 60; and gestation, 114; and litter size, 119; in Montana, 89; and parasites, 100, 101; and plant species, 85; predation on, 80, 81; radio-tracking of, 214; research on, 210; running by, 44; and shelter, 84; size of, 30; speed and endurance of, 46; swimming by, 48; and vocalization, 75; in winter, 98, 99; and young, 124
rabbit, riverine, 13, 22, 23, 84, 91; and distribution, 93; as endangered, 172, 174, 194; and gestation, 114; and lentivirus, 153; nests of, 116; as prey of cats, 193
rabbit, swamp, 20, 31, 78, 84, 171; and chin mark, 78; coloration of, 52; fur of, 61; and gestation, 114; mating by, 65; and mutual affection, 67; and nest, 115–16; size of, 30; swimming by, 48; and vocalization, 74, 75
rabbit, volcano, 12, 23, 93; and climate change, 179; as endangered, 174; and fighting, 70; fur of, 50, 54, 59, 61; and gestation, 114; and litter size, 119; in Mexico, 89; and parasites, 100; predation on, 80; size of, 30; and vocalization, 75
rabbit drives, 167, 168
rabbit farming, 188–89. See also food, rabbit as
rabbit felt, 188
rabbits, 6; distinguishing characteristics of, 3, 4–5, 6–7; evolution of, 19–28; kinds of, 8–12; species of, 5–6, 8, 9, 12–14

rabbits, types of: African rock, 50, 84, 211; albino domestic, 62; Annamite striped, 9, 14, 51, 184, 211; Angora, 145, 147, 151, 188; Asian striped, 13; black European, 52, 53; bristly, 5, 13, 22, 23, 114, 174; brush, 72, 90, 102, 114, 120; Bunyoro, 13, 30, 81, 91, 114, 116, 184, 211; Columbia Basin pygmy, 174, 175, 182; grass, 12; Himalayan, 63; Holland Lop, 145; hutch, 148; lop-eared, 147; Lower Keys marsh, 173, 175, 180, 183, 192, 193, 202, 210; marsh, 48, 51, 52, 75, 81, 114, 116, 173; Mini Rex, 145; Natal red rock, 92, 218; Netherland Dwarf, 145; New Zealand white, 151–52; red rock, 114, 116; riparian brush, 173, 174, 175, 193; rock, 9, 13, 14, 22, 23, 80, 81, 91; San Jose brush, 92–93; silver, 146; striped, 11, 22, 23; Sumatran striped, 9, 14, 30, 51, 174, 211; Venezuelan lowland, 9, 114, 119. See also rabbit, Amami; rabbit, domestic; rabbit, European; rabbit, pgymy; rabbit, riverine; rabbit, swamp; rabbit, volcano
raccoons, 80–81, 130, 190
radio-tracking, 213, 214
rats, 1, 36–37, 38, 148, 152, 163
ravens, 106
religion, 196–200
reproduction. See mating and reproduction
roads, 190–92, 194
rodents, 5, 6, 11, 20; coloration of, 49; diets of, 38; and disease, 102; hopping of, 46; and lagomorphs, 1–2, 23; and scent marking, 78; size of, 29; teeth of, 3, 38
Romans, ancient, 146, 186, 189, 199
Romerolagus, 12
running, 3, 4, 44, 53, 54, 72, 73
Russia, 170

sagebrush, 12, 99, 134, 178–79, 180, 182
Schlemmer, Max, 161
secretions, and communication, 77–78
seeds, dispersion of, 105, 110
shrews, 20, 29
Shakespeare, William, 201, 202
Siberia, 81, 82, 94
Sitwell, Edith, 203
skeletons, 5; growth rate of, 126, 127
skulls, 3, 4, 5, 45, 128
skunks, 80–81, 130
sleep, 39–40, 87
smell, sense of, 3, 35–36
snakes, 15, 81, 130
sociality, 7, 64–70, 67–68, 71, 73, 74–75, 110
South Africa, 93, 174, 193, 194
South America, xiii, 13, 14, 20, 24, 25, 82, 91, 94, 186
Spain, 93, 103, 104, 174, 186, 191, 209; Doñana National Park, xiii, 86
squirrels, 15–16, 17, 81, 130, 190, 191

stress, 52, 69–70, 106, 137–38, 144
swimming, 48
Sylvilagus, 6, 12–13, 94, 114

tails, 4, 42–43, 44, 53, 54
tapetis, 9, 20, 30, 90, 102, 114, 119, 153
teeth, 1–2, 3, 24, 135–36
territory, 64, 69, 84; and chin mark, 78; and fighting, 69, 70; and home range, 85; marking of, 77; and nest, 117; and paternity, 121; and vocalization, 75, 76
Tibet, 15, 80, 166
Tibetan Plateau, 59, 107–9, 178
ticks, 100, 101, 170
trickster, xiii, 201

United States, 12, 13, 89, 90, 94, 186, 193. *See also individual states*
urine, 36, 77, 79, 111–12, 113, 144
U.S. Endangered Species List, 173, 178

Venezuela, 9
vocalization, 6, 71, 74–76, 77
vultures, 82, 167

Washington (state), 12, 173
water, 95, 97, 138–40, 141
Watership Down (Adams), 202
weasels, 15, 56, 75, 79, 80, 82, 108
World Lagomorph Society, 209, 215
World Rabbit Science Association, 208
wolves, 15, 80, 108, 109

Xenophon, 186, 199

young, 6, 71; altricial, 6, 40, 115, 119; care for, 65–67, 76, 79, 122–26, 154–55; and chewing, 136; as food, 185; growth rate of, 126–28; huddling by, 124, 125; and infanticide, 71; and life span, 129, 130; newborn, 6, 61, 100, 114, 119, 120, 130, 147, 199; and nursing, 3, 122–24, 125, 126; and play, 73, 74; precocial, 40, 114, 115, 116–17, 119, 124, 155; as prey of cats, 193; and roads, 192; and siblings, 123, 124, 125; and vocalization, 75, 76; weaning of, 124, 125